THE BIRTH OF LANGUAGE

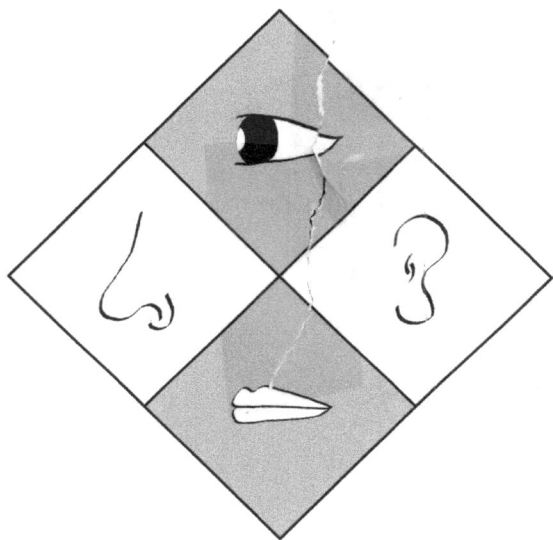

ON THE ORIGIN OF LANGUAGE
AND WHAT IT REVEALS ABOUT
THE FUTURE OF OUR SPECIES

MIN K. KIM

THE BIRTH OF LANGUAGE

Copyright © 2020 Min K. Kim

www.birthoflanguage.com

Cover photo by Emmanuel Keller

ISBN 978-1-950-13223-2

1st edition

CONTENTS

PREFACE

Almost a quarter of a century ago, I wondered if I will ever be able to speak and think in English one day. My parents and I had just moved to the United States from the other side of the planet. As a young boy, I had never been exposed to the English language. Being born and raised in Seoul, the only language I knew was Korean. My brain did not recognize English sounds at all. But everything changed one day in 1995. I started attending school in the United States. For the first time in my life since I was a little baby, I no longer had language. I could not speak or understand what anyone said. It was nothing short of a total nightmare.

I still remember the first day of school in my new country. For some peculiar reason, the school counselor thought Spanish should be my very first class. Having interacted with me, she was fully aware that I could not speak any English. But Spanish was just too important for me to miss even though that meant I had to learn it in a language that I did not speak. Fortunately, I was able to switch to a different subject. But that experience gave me a fear of learning Spanish and I never took Spanish ever again. But my English has gotten better. Much better. In fact, I even had the audacity to write this book in English. I just needed a quarter of a century.

There is a somewhat similar story that took place in 1970 involving a feral child known as Genie. She was born and raised in the United States, but her parents kept her isolated in a room at all times. When government authorities found her, Genie was about 13 and a half years old, almost exactly the same age I was when I started learning English for the first time. But Genie never really got to learn to speak properly. With intensive training, she was able to memorize many words. However, her English grammar could not fully develop despite all the efforts given by various researchers and linguists. Why was I able to learn English eventually but not Genie? Having exposure to English began at almost exactly the same age for both of us. Of course, Genie was a feral child who was most likely tortured, abused, and neglected by her parents. But what if she was rescued when she was much younger? Many feral children who were found while they were infants went on to acquire language just like other children. Some believe humans became talking apes gradually over time due to our high level of intelligence.

Others argue it was simply a mutation in our brains that gave us language rather instantaneously. Which theory is right? What did really happen that separated us from all other species by a wide margin? Not only are these intriguing questions, but they are also essential to understanding what we really are.

So I decided to take on this bold assignment and attempt to find the answers to these questions. Throughout this book, I will make claims that will not only paint a picture of our past, but also will let us see what lies ahead in the future for our species. In Chapter 1, we take an imaginary time machine and travel back to millions of years ago in order to find out how humans might have evolved before humans had language. We also discuss Noam Chomsky's theory of the innateness of language among other topics that are currently not so popular. In Chapter 2, we take a look at the human brain and how it might be utilizing language. We also examine human intelligence in comparison to other species such as birds and chimpanzees. In Chapter 3, language is analyzed in depth in order to get a better understanding of how it works and whether it might be related to math and music. In Chapter 4, we finish the book by discussing how to plan for the future and how we can survive as a species with such enormous power that we possess.

Lastly, before we begin this journey, I would like to ask you to bear with me if my writing is not as comprehensible as you hope it to be. After all, English is my second language and my brain still does not utilize it as my mother tongue. The only other option I have is to write this book in Korean since I do not speak any other language, especially Spanish.

M.K.
Fall 2019

THE BIRTH OF

LANGUAGE

THE EVOLUTION

"And God said, 'Let there be light'
and there was light."

Book of Genesis 1:3

Once upon a time, there was a big family lived in Africa. We know for certain they existed, even though we hardly know anything about them. Our existence is the evidence. It was roughly ten million years ago when the family went separate ways. For the first time in history, three different groups—Pan, Go, and Homo—were born. Luckily, all of them have survived for millions of years to this day. The Pan family became chimpanzees and the Go family became gorillas. Last but not least, the family of Homo became modern humans, the most advanced species on a planet called Earth.

Some people like to argue we are not *that* different from chimps and gorillas. After all, we share the same ancestors. At the same time, people also think humans are very different from one another. But it turns out variance is rather difficult to find within the group of Homo sapiens. Our DNA shows us we are basically distant relatives. More importantly, we all have the ability to comprehend language, mathematics, and music. That makes us rather unique in the animal kingdom. We stand out from all other species of animals, including monkeys and apes.

So what happened exactly? What was it that gave us the ability to comprehend the world and made us wonder about our own existence? Well, let's start from the very beginning.

Figure 1-1. The March of Progress
Our ancestors slowly began to walk upright over time. The cause of bipedalism is still being debated.

The Forbidden Word

Alister Hardy, a marine biologist, wondered why the human body is wrapped with a fat layer. This is clearly a trait of marine mammals. Terrestrial mammals, on the other hand, have fat around their organs and not around their skins. This and other human traits made Hardy hypothesize our ancestors might have been partially aquatic in the past. His hypothesis, known as the aquatic ape theory, says that the early human species came down from trees and settled down near big streams of water. However, academia were quick to reject the idea and declared it pure nonsense.

Then came a writer named Elaine Morgan, who came to Hardy's defense. She favored the aquatic ape hypothesis over the more traditional savanna hypothesis, which claims human ancestors moved away from trees and onto savannas. Morgan went on to propagate Hardy's "aquatic theory" until her death in 2013.

If Hardy and Morgan did not use the word *aquatic* and called the hypothesis *semiaquatic* or *waterside* instead, would people have been more acceptable of the idea? Hardy did not claim our ancestors were marine mammals like whales. He barely contended the idea that our ancestors spent some time of the day in water. Hesitant to put his reputation on the line, Hardy called the hypothesis a mere speculation. (Morgan, on the other hand, was a lot more unyielding.)

So far, there has not been any concrete proof that can declare Hardy's theory one way or another. The jury is still out, even though not many are willing to tackle it. But the importance of this conjecture is clearly indisputable. If the "aquatic theory" gets accepted, we have no choice but to rewrite history.

Figure 1-2. Survival of the Tallest
Humans have evolved to be great swimmers. But why?

Terrestrial vs. Aquatic

Biology teachers like to tell their students bipedalism is the hallmark of human evolution. After all, it is hard to ignore the fact we walk on two feet while apes are quadrupedal. (However, they can walk on two feet for short periods of time.) But bipedalism is not necessarily a human trait. Birds, kangaroos, dinosaurs, and a few other species belong in the same club. One thing for sure, bipedalism was very beneficial for our early ancestors. Some say being tall made them spot predators more easily. But that does not explain why most mammals–prey or predators–have not also become bipedal. A more likely reason why we slowly began to stand vertically was to spend more time under water.

Take human babies for an example. In nine months, they turn out to be the chubbiest newborn creatures on Earth. But this is a rare phenomenon among animal species. Almost all animal babies are born slim, including newly-born chimpanzees. Compared to our own plump creations, chimp babies are quite slim. It is no wonder human babies are incapable of doing any task for at least first 12 months. Since they can hardly move, they just lie on their backs and suck on their thumbs. On the other hand, baby chimps learn to walk almost right away after they are born, which seems like a good idea and much better for survival.

But there is one advantage for having a double-digit percentage of body fat; this way, the body floats on water. The density of fat ($\sim$0.9 g/cm3) is lower than the density of water ($\sim$1 g/cm3), increasing the level of buoyancy for human babies. However, the density of muscle ($\sim$1.06 g/cm3) is greater than the density of water. That means having more muscles like non-human primates would make our bodies sink. Being very muscular, it is not a surprise chimpanzees and gorillas do not float on water like humans do.

Whatever the reason may be, it is a fact that evolution changed our body density to be slightly lower than the density of water. When a human baby is placed in water facing upward, her face naturally pushes out of the water just enough for her to breathe. Even when babies are nowhere near water, they still move their limbs around as if they are swimming. The hind-leg kicking movement human babies do is not a trait we share with apes. Ape babies use their limbs to firmly grab on to their mothers' torsos. Since human bodies are mostly without hair, it would be even more beneficial for our babies to glue themselves to their mothers. But, instead, human babies kick their legs as if they are natural-born swimmers.

The likely explanation of this instinctive behavior seems to be that human babies are aquatic from birth. While babies that were not aquatic did not survive, the ones who were aquatic did. That would make us the remnants of this evolutionary trait. Because, after all, today's humans are just as aquatic as our ancestors were. (Technically speaking, it would be more accurate to describe Homo sapiens as semiaquatic and not fully aquatic. A person who can swim, by definition of the word, is semiaquatic.)

On the surface, it may seem as if this topic bears no significance when it comes to the origin of language. But the next event that potentially took place might say otherwise.

The Real Hallmark

Once we became "water-friendly" apes, we no longer had to depend solely on fruit and vegetables for food. We could now digest meat as well. Fossil records show our ancestors enjoyed devouring fish, water vegetations, sea orchins, alligators, and hippopotami. But being carnivores was not easy. Now the food had legs.

It was one thing for our ancestors to pick fruit from trees, but catching fish was another matter. The archaic human body was designed for living on and around trees. We had no fangs, claws, horns, or wings. We were also slow and weak.

When we look at it from this perspective, Homo sapiens appear be the most mediocre creatures in the world. It is especially true since we had millions of years to do something about it. But evolutionary biologists would say this is impossible. All species of animals evolve to adapt to the environment in some physical form.

Figure 1-3. Four vs. Two
Bipedalism is not necessarily better than quadrupedalism. But it did change the shape of our bodies.

The fact is, if Charles Darwin was right about his theory, the human ancestors had to evolve to gain some physical advantage. Once again, the aquatic ape theory might be the answer. Unlike other species, we developed the ability to swim and obtain food in water. Even today, people are capable of catching seafood without using sophisticated tools. (Professional female divers in South Korea, known as *haenyeo*, make a living by catching seafood using their hands and simple hand tools.)

But it probably was not enough that we became partially aquatic, which explains why our brains started to increase in size and mass. (Another explanation would be that the brain had to become larger, helping the body float in water.) Whether we were catching fish or hunting mammals on land, our intelligence had to make up for our inferior exteriors. Bipedalism and aquatism were only the stepping stones. What may have separated us from all the other species is the change of our diet that eventually turned us into highly intelligent predators. For us to become successful hunters, we had to be more intelligent. If we had stayed arboreal, it is very likely Homo sapiens would not exist today.

The Mighty Hunter

After adjusting life around big waters, our ancestors likely looked for food elsewhere once again. It could have been by choice, but more likely the environment underwent another major change and the water dried up. No longer able to find food inside water, they either had to go back to living on trees or try catching terrestrial animals. Fossil records favor the latter hypothesis. By this time, the early humans would have been more prepared to hunt down even lethal predators like leopards and hyenas. They began using tools for capturing and killing animals. If they could go after alligators and hippos, they certainly had a shot at killing big mammals on land. But could they possibly have done it without having language? Wouldn't it make more sense for them to have communication if they hunted together?

Then again, we have seen non-human primates catch monkeys in groups without saying a word to each other. A monkey gets cornered into a certain direction by a couple of beta-male gorillas and then the alpha-male catches it from the other side. When the monkey is caught, its meat is shared among the hunting members. They achieve to do this without having any communication or language.

If humans started developing language millions of years ago as they became mighty hunters, it would not have taken much time for us to reach where we are today. We could easily take out millions of years from our evolution timeline. However, fossil records do not indicate any type of significant advancement from this time period. The early humans, if we may call them that, merely used simple tools to catch food. They were still very much primitive. But their bodies and brains slowly began to change.

The Transformation

Since the arboreal days, the body of hominids has gone through various changes. Seeing a cat for the first time, an infant might point to the cat and says "dog." In his brain, there is enough similarity between dogs and cats that he may think the two animals are the same kind. But the infant would never confuse his father with a chimpanzee, no matter how hairy his father may be. The human body no longer resembles anything else.

Even if our early ancestors had complete body hair, they still looked quite different from chimps. Compared to non-human primates, the human face went through major changes. The nose is a prime example. Our legs also became longer than our arms, which is not the norm for primates. Chimps and gorillas have hands and feet that look virtually the same. But the shapes of our hands and feet do not match anymore.

We also became much taller as we began to walk upright. Pound for pound, humans are taller than any other primate group. Some may say it was advantageous for our ancestors to stand tall on the ground. If this were true, many other species of terrestrial animals in the same habitat would have gotten taller as well. Not only has this not occurred, almost all other land species remain quadrupedal as well. It is more likely that the cause for our height growth has to do with our time spent in water; the taller the person, the more likely she gets to survive.

As height increased, so did the size of the brain. Today's human brains weigh about three times more than brains of chimps or gorillas. But the ratio does not stay the same when it comes to body weight. Gorillas weigh far more than humans on average, even though our brains are much larger than their brains. It appears that—as we became more and more predatory–our brains grew in size and weight. Predators and carnivores like bears tend to have bigger brains per given body size than prey and herbivores like buffaloes. The reason may seem almost too obvious. Clearly, having more brain power would be advantageous for hunting animals. Then again, other species of mammals do not walk around with big noggins like we do, even though they also hunt for food. Perhaps it is also possible that the brain had to increase in size for the entire head to balloon so that the face will break out from the surface of water in order for the baby to breathe.

No matter what the cause was, once meat became part of our diet, we started having enough nutrients to develop more brain cells. But this would mean our brains did not necessarily increase in size for us to create language. Instead, the brain became bigger so we could survive and adapt to our environment. Fossil findings show that the brain size grew steadily over time. There was no exponential growth.

If we had developed language very early in our evolution timeline, we probably did not need such big brains. It would also suggest other species of animals might have developed language with their small brains. But that is not what happened. Even species with big brains such as elephants, whales, and dolphins do not show any signs of having language like humans do. Many researchers have tried to give apes language, but they all have failed. However, some success did occur with Kanzi (bonobo) and Koko (gorilla). They did learn sign language to some extent and could communicate with humans in simple forms. But one thing became very clear; other animals cannot utilize language like we can. Apes and non-human primates, as lovely as they may be, are nothing like Homo sapiens when it comes to cognition.

The Sullivans vs. The Kellers

At some point in our past, the people were endowed with the gift of language. A few inguists–people who study the properties of language–tried to figure out how and when language came about. One notable linguist is a man named Avram Noam Chomsky (better known as Noam Chomsky). He is considered by many to be an intellectual of the modern era. He claims language is rather a recent phenomenon than people might suspect. It just happened one day to our species and it was so useful it stayed. Many people–linguists and others–are not convinced of this hypothesis. They believe it is more likely language slowly became what it is today over a long period of time. In other words, natural selection helped our ancestors evolve into talking apes. Clearly, only one theory can be right. The creation of language could not have occurred suddenly and gradually. So let's take a look at each hypothesis and see which one is more likely to be true.

First, we will start with the theory that claims humans gradually evolved to acquire language, which I will call the Sullivan hypothesis. We can call the theory of sudden language installation in humans the Keller hypothesis. Now we can address the two sides as the Sullivans and the Kellers. (This way, we can avoid addressing people as "Chomskyans" and "anti-Chomskyans.")

Let's imagine a world where the Sullivans are right. At sometime in the past, the human ancestors decided to communicate with each other using hand signs or simple sounds made with our vocal cords. They put two or three words together to form basic sentences. The communication mechanism at this time would be too primitive for them to create more complex expressions or thoughts. But a few individuals might have been slightly better than others at combining words together. If this helped them to be more successful in their daily endeavors, the language capacity would have gotten better over time for our species. This theory is based on natural selection and the theory of evolution. Not only that, fossil records also appear to support the Sullivans' assertion. There is no evidence that the human brain increased exponentially in size in a short duration of time. Fossils tells us it was actually the opposite; our brains evolved gradually over hundreds of thousands of years. Thus it would make sense for language to have developed over a long period of time as well.

The Kellers, like Chomsky, firmly believe language was created due to a few or even a single mutation. A genetic mutation in one individual's DNA enabled her or him to have the power of language all of a sudden. Only then, evolution came into play and we ended up as noisy chatterboxes over time. This means there was little to no continuous development of language; we just received language one random day as if some superbeing decided to give it to us. Charles Darwin likely would have been more of a Sullivan than a Keller. Even to a layman, the Keller hypothesis is rather a hard pill to swallow. Then again, why does someone like Chomsky–a highly-regarded thinker–believe it to be true?

The Language Organ

According to Chomsky, language functions as an organ like the heart or the eye inside the brain. Neuroscientists agree that this seems to be true. A few key areas were discovered in the human brain that are dedicated to language utilization. Broca's area, named after Paul Broca, is located in the left hemisphere. This region is responsible for putting words together or creating sentences. Another important region in the same hemisphere is Wernicke's area, named after Carl Wernicke. This part is generally responsible for analyzing language data. If there are problems with either Broca's area or Wernicke's area, one will not be able to speak or sign normally.

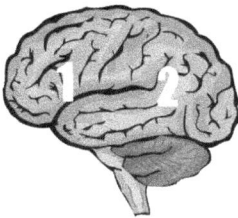

Figure 1-4. Broca's Area and Wernicke's Area
In general, Broca's area (1) is responsible for sentence formation whereas Wernicke's area (2) is responsible for language comprehension. They are usually located in the left hemisphere.

The Sullivans would argue these regions positioned themselves slowly into our brains over a long time. The Kellers, like Chomsky, believe the areas were either created or reshaped for language instantly. Whatever the truth is, one thing does seem certain; the brain must have the "language organ" for it to utilize language. Neither having large brains nor being exposed to language is sufficient. One must be given the necessary apparatus for him to generate and process complex thoughts. In other words, language appears to be innate.

The Critical Period

When my parents and I moved to the United States, we ran into a few other immigrant families from South Korea. Most of the children, like myself, lacked even basic English skills. We all had to learn the language from scratch. It was an even playing field. But, after a while, I began to notice something pecular. The younger sibling of each family was learning English more quickly than the older sibling. Being younger—especially for the preteens—was apparently a big advantage.

They were not only learning English faster than the older kids, but they were also better at it. The older children like myself just could not keep up with them even when it came to something as simple as word pronunciations. And they were doing it with minimal effort, which did not sit well with their older siblings. This was always the case with all the immigrant families. However, no one knew exactly why.

Little did I know at the time that this phenomenon is known as the critical period hypothesis. It claims a native language or first language (L1) cannot be acquired later in life once the person reaches a critical period. That means a child must be exposed to language very early in her life. If she is not exposed to language by the time she is 13 years old, the chances of her developing language is almost none. A second language (L2) also becomes more and more difficult to acquire as one reaches puberty and adolescence. Language learning is a ticking time bomb; one must act fast before he runs out of time. The sooner one is exposed to language, the better.

When I became an English tutor, this hypothesis showed its face once again. The mothers of little children marveled at my "teaching skills" whereas the mothers of teenage students were not as impressed. Little did they know, it was actually the children's brains that were doing most of the work. All I did was simply give my students exposure to English the right way. Then here is a dilemma; if we are capable of learning language, why does it get more difficult as we age? A 40-year-old or a 60-year-old is as capable of learning history or science as a 10-year-old, if not more capable. But language seems to have a time limit unlike other academic subjects.

No matter whether the Sullivans are right or the Kellers are, we must come to the conclusion that language is indeed innate. Humans are born with the ability to acquire language from birth. Broca's and Wernicke's areas are prime examples. But even the eye loses its abililty to see if it is not exposed to light early in life. Apparently, language is no different. If I happened to be just a few years older when I arrived in the U.S., this book would have been written in Korean.

The Neanderthal Paradox

Before humans began migrating out of Africa around 60,000 to 200,000 years ago, another human species made the same journey. According to researchers, Homo neanderthalensis– more commonly known as Neanderthals–began exploring other continents at least 100,000 years ago. Before Neanderthals went extinct around 40,000 years ago, the subhuman species became successful dwellers throughout Europe and other places.

When scientists extracted the DNA of Neanderthals, they realized there was little to no difference between them and Homo sapiens. The Neanderthals' phenotypes (visible characteristics) and genotypes (genetic makeups) both indicate they were pretty much human, even by today's standard. In fact, many people living in the 21st century still carry a few percentages of Neanderthal genes in their DNA. This means Homo sapiens and Neanderthals used to dwell together outside of Africa. We also discovered that Denisovans, another subhuman species, had coexisted with both Neanderthals and Homo sapiens sometime in the past. This is why most people from outside of sub-Saharan Africa still hold on to their ancestors' archaic genes, including the author. That is why, in a sense, many of us are not entirely humans.

Fossils, stone tools, and other remains of the past reveal Neanderthals were quite sophisticated. The big question is, did they also have language? The fact that they wore clothes and made tools might seem to indicate the presence of language in their brains. Research shows that their vocal cord anatomy would have allowed them speak like we do today, although their voices might have sounded slightly different from human voices. Many researchers and scientists say they believe Neanderthals likely had language. Taking a look at a reconstruction figure of a Neanderthal gives us an intuitive sense that they must have had language like the rest of our ancestors. It is very difficult to imagine otherwise when Neanderthals looked so much like us.

Unfortunately, this theory raises a few problems. If Neanderthals—and possibly Denisovans—could have spoken, then that would indicate that the first speakers were our common ancestors before the migration out of Africa. But these early humans split into Homo sapiens, Neanderthals, Denisovans, and perhaps others somewhere between 400,000 and a million years ago. It is quite puzzling how none of these groups did anything for such a long time with language in their brains. A more likely possibility is that Neanderthals developed language on their own after the separation. The Sullivans might say this is very much a possibility. If language gradually evolved in humans, it also could have happened to other human species like Neanderthals.

No matter what the truth is, there is little doubt that Neanderthals were intelligent beings unlike how they are portrayed in most documentaries. They successfully migrated into other places on Earth and survived in harsh climates for many years. They only went extinct when Homo sapiens from Africa knocked on their front doors around 40,000 years ago. Some people claim our ancestors were so superior they simply eradicated Neanderthals from the planet. But our genes show they got along just fine with other groups of humans. Even if Homo sapiens behaved in a hostile and threatening manner, Neanderthals easily could have moved to other places and continued to survive. The world was a big place back then, especially since the only means of transportation was walking barefoot. The mistake Neanderthals made is more of the opposite; they were too friendly and unwary of the newcomers. The most likely cause for Neanderthals' demise (other than a volcanic eruption) has to do with diseases Homo sapiens brought from Africa.

When Europeans began settling in North America in early 16th century, it was diseases more than anything else that killed nearly every indigeneous person living in the continent. Having no immunity for what the Europeans were carrying had to be too lethal. The population of Native Americans plummeted very quickly as soon as settlers arrived from Europe. If members within the same species could spread such deadly diseases for others, it is more than a possibility that Homo sapiens were carrying germs that eventually wiped out Neanderthals.

As of today, there is no convincing evidence that says Neanderthals were significantly less intelligent than Homo sapiens. Fossil records show Neanderthals actually had as much as 50% more brain mass than today's modern humans. Having bigger brains does not necessarily equate to having higher levels of intelligence. However, it is still a difficult proposition to assert that Neanderthals went extinct simply because they were inferior cognitively compared to the Homo. If the history of humans tells us anything, then it is far more likely for a superior human species to enslave the others. Mass extinctions, on the other hand, are almost always caused by Nature or indirect human behavior.

Chomsky's Postdiction

So it might appear that Neanderthals were lingual, especially since their brains were enormous compared to ours. But this theory does not mesh well with the current evolution timeline. Chomsky thinks only Homo sapiens could have possessed language starting around 80,000 to 60,000 years ago as the earliest date. He believes this because Neanderthals and Homo sapiens were separated at a much earlier time period. That means the only way the big-brained anthropoids could have had language is the possibility that they also had the language mutation occurring in one or more individuals within their group. But the chances of two different species having the exact same mutation around the same time are not that high. Then again, how can one explain Neanderthals' ability to make clothes, use tools, and perhaps even be able to create cave paintings? Is it possible they could have done all of that without having language? The answer seems to be yes.

Orangutans collect leaves and put them on their heads when it rains. Neanderthals being far superior intellectually than orangutans, it would not be such a stretch that they made clothes when orangutans are capable of making umbrellas. Gorillas and chimpanzees are also known to use tools such as rocks and sticks. Elephants can learn to paint. Beavers can make dams. Birds build nests. Aside from symbolic or abstract creations, language does not appear to be a necessity for the possession of creativity.

If Chomsky's postdiction about Homo sapiens being the only lingual species proves to be true, then the Kellers will claim victory over the Sullivans. However, there is not enough evidence gathered so far that indicates whether Neanderthals certainly had language or not. At the same time, Neanderthals also seem to have evolved to possess the capability of making sounds using their vocal cords. Thus, it should not be a big surprise if they communicated with one another using simple expressions. (It also raises the possibility that they had music just like we do today.) Non-human primates can be taught to put words together in rudimentary ways (e.g. "Koko love," "There apple there"). But these animals seem to lack grammar that is required to form complex sentences or thoughts. If Neanderthals were missing the "language organ," that would mean they had no grammar or syntax, which is crucial for language utilization.

Scientists tells us Neanderthals were 99.7% identical to modern humans, genetically speaking. That would mean around 70 genes are different between the two species. Chomsky believes humans acquired language most likely due to a single gene mutation. That would be just one mutation in one gene out of 70. His math checks out. For the Sullivans, however, numbers present a conundrum. If language gradually evolved into the human genome, then it does not explain why the same did not happen to Neanderthals with larger brains. It would also push back the timeline for the development of language from around 100,000 years to half a million years ago. During this time period, there was little difference between Homo sapiens and Neanderthals, if any. Even today, the difference amounts to only 0.3%. In other words, our human ancestors could not have been far superior than other subhuman species. This suggests Neanderthals likely developed language like Homo sapiens did, which does not explain why we have not found any symbolic activity from Neanderthals.

If we were to push back the timeline of language development in humans even further, we run into two different problems. One, it makes us wonder why we did not make any progress as a species for almost a million years. With language in our possession, we should have started farming hundreds of thousands of years ago. But farming only began a little over 20,000 years ago. It is quite difficult to imagine the idea of agriculture took that much time to develop if language was present. Another issue with this scenario is that it puts us back into an era where Neanderthals and Homo sapiens were the same species, which means Neanderthals must have had language, too.

That would make Homo sapiens no different from them. Well, can it be true?

Different Species After All

The word *species* is commonly defined as a group of organisms that are genetically compatible enough to produce healthy offsprings with no major side effects. For instance, it is true that a female horse and a male donkey can give birth to a mule. But the mule is likely to have problems producing offsprings. The same is true for hinnies, which are hybrids of male horses and female donkeys mating with one other. If horses and donkeys belong to the same species, reproduction would not be a big issue.

We know for a fact Homo sapiens and Neanderthals made babies together. The proof is in our DNA. This might tell us the two groups were perhaps the same species with only minuscule differences. However, scientists found out today's humans do not contain any Neanderthal DNA in their Y chromosomes. The Y chromosome is only passed down from male to male. This means male Neanderthals and female humans did not or could not create healthy boys. Unless the baby's sex was female, the pregnancy would either end up in a miscarriage or the baby would not survive for long after birth.

This piece of evidence indicates Neanderthals and Homo sapiens are two different species indeed, like horses and donkeys. It also suggests most female humans of today–excluding women in sub-Saharan Africa–should have a few more Neanderthal genes then the male counterparts. By a small percentage point, women today are more Neanderthal than men are on average, which is quite an irony since men's behavior normally is considered a lot more "Neanderthal." Violent crimes, especially murder, are mostly committed by men. Experts blame the presence of the Y chromosome in men as the culprit, although hormones could also play a role.

The fact that interbreeding took place between Homo sapiens (H.) and Neanderthals (N.) gives us four possibilities of genetic combinations between the female (f) and the male (m):

[1] H. (f) × H. (m) = 100% Homo sapiens
[2] N. (f) × N. (m) = 100% Neanderthal
[3] H. (f) × N. (m) = 50% each
[4] N. (f) × H. (m) = 50% each

Offsprings carry Neanderthal genes in [2], [3], and [4]:

[1] Boys, girls = 0% N. genes
[2] ~~Boys, girls~~ = 100% N. genes
[3] ~~Boys~~, girls = 50% N. genes
[4] Boys, girls = 50% N. genes

But boys in [3] are stillborns. Also, boys and girls in [2] are 100% Neanderthals, making them irrelevant. So the ratio between boys and girls is two to three, resulting in a higher number of females overall. If this actually took place, it suggests there were more women in existence compared to men at some point in our past. It also implies women who mated with Neanderthals were at a disadvantage since they would not produce as many healthy babies. But women who copulated with Homo sapiens would produce perfectly healthy boys and girls. As a result, Neanderthal genes slowly faded away in our species.

Language and Intelligence

So having a few percentages of Neanderthal DNA turns out to be no big deal. It is part of our lineage, thanks to everyone getting along with each other some eons ago. Perhaps people of today can learn a thing or two from our ancestors. We should also ask ourselves whether humans would be more intelligent than other species if we do not have language.

The Sullivans might say we started speaking to each other because we were smart enough to do so. But how smart could we have been if we had no language to begin with? It is true we developed big, sophisticated brains. However, some other mammals have even larger brains than ours. We have enough evidence to believe our ancestors used stone tools and fire. But the usage of such primitive technologies dates back millions and millions of years, way long before language showed its face for the first time. Considering the fact that even tiny insects like ants and bees have some forms of communication, it would not be a big suprise if our species also communicated with each other without having the language faculty. Also, while our species made stone tools, birds with their tiny brains made nests. (It is highly doubtful whether an average person today can beat a bird at making a nest in terms of speed and quality.) Scientifically speaking, there is no evidence that our early ancestors were intelligent enough to create language from scratch.

Even before language, we were already accomplished hunters. Otherwise, an archaic subhuman species such as Homo erectus–who clearly did not show any sign of having language–likely could not have proliferated for over a million years. We, as a species, had proven that we could survive in any environment under any condition. It seems like we were already smart enough.

The Sound of Music

One of the most important questions regarding the origin of language has to do with the origin of music. Did we have music before we began to speak or was it the other way around? Perhaps music and language appeared about at the same time.

Many people have attempted to find similarities between language and music to see if they are essentially the same. For one, they both use sound as a medium. They are also uniquely human traits. However, unlike language, music seems to be totally pointless. It serves no apparent purpose to the naked eye. Let's imagine a bird species exists that is emotionally moved by the presence of chameleons. The birds hop up and down on tree branches happily as they watch chameleons change their colors. In fact, the avians do nothing but sit on trees all day long and watch chameleons put on a show. But what would be the benefit of such behavior? Most likely, it would be a total waste of time. A species of this kind might have existed, but it would not have lasted long. Now let us imagine an early species of humans that somehow evolved to love music. They would sing every day or listen to others sing. What purpose would that serve? Making melodic or rhythmic sounds repeatedly would not have given them food, water, or protection. The only thing music supposedly gave people was pleasure or goosebumps. That syndrome eventually turned them into musical beings. Only then music became useful for something, sex. According to Charles Darwin, "musical notes and rhythm were first acquired by the male or female progenitors of mankind for the sake of charming the opposite sex."

It is somewhat hard to argue that this hypothesis is completely false. Birds, our long-time neighbors in Nature, sing for the same purpose. In most bird species, it is male birds that sing to attract the female. The same can be said about Homo sapiens. Generally speaking, women display more sexual attraction toward men who sing. There is a reason why boys in a band are very popular with girls. And it is usually the lead singer that steals girls' hearts. But the same is not necessarily true for the opposite sex. Boys and men tend to care more about visual attraction than anything else. A woman does not necessarily have to carry a tune as long as she *looks* healthy and is neither too young nor too old to bear children.

It seems like Neanderthals had music, too. According to various researchers, a couple of flutes made out of animal bones were discovered that were likely made and used by Neanderthals. If this turns out to be true, then it is very much possible the common ancestors of Homo sapiens and Neanderthals also had music. Then the origin of music would have to be dated back to at least hundreds of thousands of years. It would also explain why we have evolved to make complex sounds using our vocal cords. Once language arrive, we then began using the instrument to convey meaning.

When Pigs Talk

If music does predate language, that might explain why no other species of animals have developed speech. According to the Sullivan hypothesis, even animals can evolve to develop language step by step because that is exactly what happened to us. But we see no evidence of this theory in reality; only humans possess language with no exception.

On the other hand, the Kellers can claim the current knowledge of our past supports their hypothesis. Their theory postulates an apparatus for making speech already existed. This is because our vocal cords had to evolve for music way before we started speaking. Coincidentally, our hands also happened to be dexterous for verbal communication. Once we had the language mutation, we then could have either spoken or signed. Chomsky calls this *externalization of language* ("E-language"). But he argues, before language can be externalized, it must be *internalized* inside the brain ("I-language") first. In other words, without I-language, there is no E-language. Language must be in our brains before words can come out of our mouths. Chomsky believes the reason why we are able to think and talk in words is because something gave us I-language. Many

Sullivans argue profusely against this idea, declaring it pure nonsense. They believe it was humans that invented E-language because we were bright. But it is rather unclear how smart our ancestors actually were for them to create language without any assistance like I-language. For hominids and other animals to develop language on their own, they either have to be very intelligent (the Sullivan hypothesis) or they have to have something like music that can be a precursor to language (the Keller hypothesis).

Again, this raises the question of whether Neanderthals had music in their brains or not. Because music appears to be a much older trait than language, Neanderthals possibly did enjoy music like us humans. If this is true, the language gene could have happened to Homo neanderthalensis and not to Homo sapiens, potentially speaking. Had the language mutation resulted in a Neanderthal, our history would be nothing more than a work of fiction. It is uncertain whether humans with smaller brains could have competed against talking Neanderthals with larger brains. With the power of language in their hands, Neanderthals could have been the last ones standing.

Our Galileo in the Making

Noam Chomsky became a student of language quite early in his life. While he was still a teenager, he decided to major in linguistics after meeting a linguist named Zellig S. Harris. Little did Chomsky know he would go on to revolutionize the field for the next several decades. But he has been more of a political dissident for many, especially during the turmoils of the Vietnam War era. Through his books and speaking events, Chomsky created many friends but also just as many foes for his ingenious linguistic ideas and his politicial views regarding the U.S. government.

For someone who studies and teaches about language, Chomsky's words are rather hard to swallow; whether because his ideas are seldom self-explanatory or because he tells the embarrassing truth about the military-industrial complex. His nomenclature for his linguistic theories gives the impression that he might be more of a computer programmer at heart. Monikers such as universal grammar (UG), transformational generative grammar, language acquisition device (LAD), E-language, I-language, deep structure, and Merge are seldom intelligible. His words and ideas are notoriously abstract, casting a shadow over the brilliant mind of his and putting many people in oblivion whenever they attempt to pick his brain.

If someone asked Chomsky to describe how one can apply shampoo on his hair, it might have been like the following:

> First, you have to define what shampoo really is. Unlike what the name might suggest, it's not a type of excretion at all (shrugs). Frankly, it sounds like a complete misnomer when it's basically a composition of chemicals designed for hair cleansing. It has absolutely nothing to do with bodily secretions whatsoever. If I could name it, I would call it "viscous hair cleanser" or something like that (shrugs). Well, if you want to apply shampoo, you need to put some on the palm of your hand and apply it to your hair gently. Of course, you'd probably want to wet your hair first. Then you use your fingers to massage the hair on your scalp. Once you're done, then you can wash it off with water. Now the fascinating part of all this is, you can repeat this process an infinite number of times. This very nature of applying shampoo is so unique, I recently wrote a paper, The Property of Infinite Recursion in Shampoo Application...

Most people would go with "lather, rinse, and repeat." But Chomsky writes the way he thinks and he clearly does not think like anyone else. Chomsky also seems to care little about being convincing as long as he believes he is right. Many other scientists also suffer from this condition, but they get to express their thoughts in mathematics. Galileo Galilei, Albert Einstein, and many other brilliant scientific minds had math as their second tongue. Their theories were much more persuasive since they were presented with numbers and equations, making them precise and testable. But mathematics and linguitics are like oil and water. They do not mix. It is like describing colors with smells. Chomsky and other linguists are limited in their pursuit of finding the origin of language, partially because they depend so much on language itself. It is also rather difficult to convey ideas objectively without using an independent medium like math.

Figure 1-5. No Solidarity for Chomsky
Can an unpopular opinion turn out to be true? Galileo thought so.

But does that mean Chomsky is wrong or is he another Galileo in the making? So far, no one has disproved his linguistic theories. Some merely argue that his theories are getting old, and therefore, have become obsolete. If Chomsky is wrong about the origin of language, why can't anyone–including the author of this book–prove him wrong? His critics' most common responses are either "He *must* be wrong" or "The consensus says he is *likely* wrong." These kinds of statements are even less convincing than Chomsky's metaphysical schizophasia or word salad. Well, if no one can prove him wrong, can he turn out to be right?

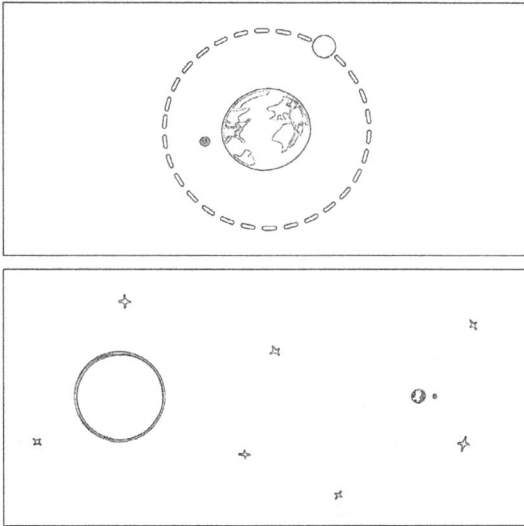

Figure 1-6. Geocentrism vs. Heliocentrism
One model made intuitive sense. But the truth belonged to a completely different model.

Chomsky claims language is innate. Humans are born with the ability to acquire language from birth. Without this trait, we will not be capable of having complex thoughts. He also believes this innate nature of language entered our DNA due of a single mutation that took place less than 100,000 years ago. Chomsky believes language has not changed form since its creation, even though its origin dates back to tens of thousands of years or more. This means language was perfect to begin with. At Winona State University in 1998, the professor of linguistics stated, "Language is remarkably well designed, that is pretty close to optimal design." If he is right, it means we will not be able to create more complex language in the future (without another mutation taking place). In 2012 at the University of Maryland, he spoke, "Remember, language hasn't evolved at all for 50,000 years since humans left Africa." He continued, "If you go back about 50,000 years before that, there really isn't any evidence there was language."

According to Chomsky, the language mutation likely happened to one individual in our species. Even though this person may not have utilized language, he or she passed down the gene and it eventually spread to the entire population, which is also known as gene fixation. Chomsky explained, "So somewhere in a very small window, something happened and nothing happened afterwards; that part we're pretty sure of." He also said, "Sign language is just another language," meaning there is no fundamental difference in the way language is externalized, whether one signs or speaks. In other words, sign languages such as ASL

(American Sign Language) are 100% pure languages, just like Romanian or Cantonese. Although hand signs are visual, neuroimaging reveals that the linguistic data is handled mostly by the left hemisphere as if the hand signs are spoken words.

Perhaps the most discombobulating claim made by Chomsky has to do with the main purpose of language. The renowned linguist asserts that the true purpose of language has nothing to do with communication. Instead, this unique cognitive feature has everything to do with thought. In Chomsky's mind, language is supposed to be an internal function for the brain. We happen to externalize it for communicating with other humans because we are social beings. But language was not never meant for social interaction. Chomsky's logic is kind of like this; our hands did not evolve for us to sign, but we can use them for signing nonetheless. In 2012, he said, "So I think it's just a big mistake to identify communication with language." Chomsky points out we spent way more time thinking to ourselves than talking to someone else. It is true our most common use of language is for thinking and not talking. Then again, we use our hands for using electronic devices like computers, even though they were evolved for climbing trees and picking fruit. Just because we have more opportunities to think to ourselves than to talk to people, it does not warrant the true nature of language one way or another. The only problem is he still might turn out to be right about, well, everything.

There are three different forms of communication; action-driven communication (ADC), emotion-driven communication (EDC), and thought-driven communication (TDC). Action-driven communication is used for urging someone to take an action (e.g. "Please sit down"). Emotion-driven communication allows us to express our emotions (e.g. "Oh my Lord!"). Thought-driven communication is used for sharing abstract thoughts and opinions with others (e.g. "I disagree with your assessment"). For ADC and EDC, language is not a requirement. Animals are fluent at communicating with each other to get what they want or express their emotional states. A honey bee can dance in order to give directions to other bees. A cat hisses under a lot of stress, not too dissimilar from the way a person curses to show anger. But animals do not make use of TDC like we do. Without language, thought-driven communication is not a possibility, at least not at an abstract level. This raises a dilemma; if we could not materialize thinking into explicit thoughts without language, then how did language evolve for thought-driven communication when thoughts did not exist in the first place?

Like other species, early human ancestors likely did not have discrete thoughts before language was present. Of course, they still had some capacity of thinking like other animals do. But making the transition from unconscious thinking to having conscious thoughts would not have been possible without language acquisition. This means the need for TDC could not have taken place before language arrived. And we already had means of utilizing ADC and EDC without having the language faculty. If this is correct, the order of progress should look like the following:

Language ▷ Thought ▷ TDC

This appears to be Chomsky's reasoning. Originally, the purpose of language was to convert the analog state of thinking into the digital state of thoughts. Only when thoughts became complex and useful enough, we developed ways to share them with other humans (e.g. ASL, Spanish, whistling). That would make language, in essence, an analog-to-digital converter. Communication is mere an expansion of this utility for our social needs and not a necessity of language.

A Quest for the Sullivans

Ongoing developments in neuroscience and genetics give us some mighty hope the search for the origin of language will likely be over in the near future. Once we find the answer to this longing question, Chomsky's legacy will be cemented for good. He could very well end up as one of the greatest intellectuals of the last several centuries. His name likely will be remembered with other great minds like Galileo, Darwin, and Einstein. But, if it turns out his linguistic views are wrong, his name will be ridiculed and then brushed aside as the Sullivans take victory.

But the Sullivans will be judged one way or another as well. What is unclear is how history will write their perspectives on the origin of language. The Sullivans include professional linguists and scholars in other academic fields. Some contend they actually agree with Chomsky to a degree whereas others vehemently dismiss the entire doctrine of Chomsky all together. Unlike the Catholic Church—which did have persuasive explanations of their own theory in geocentrism—the Sullivans have yet to develop a standard model for their theory on the origin of language.

Once a Sullivan and now a fully converted Keller, I would like to present the following 25 requests and questions for believers of the Sullivan hypothesis. If they can provide sufficient responses to most of the inquiries, then we will have another model for explaining what happened to our species a very long time ago. If the Sullivans cannot provide answers or the answers turn out to be contradictory to one another, then that would be a big hint about their theory. Here are the 25 challenges:

1. Determine or estimate when humans began to create language.

2. Determine or estimate how much time it took for language to become what it is today.

3. Is language still evolving? If not, when did it stop?

4. If language can have a range of complexity, then what would a primitive language look like? Create one to show that language can be utilized in a much simpler way.

5. If language evolved gradually, then explain why all human languages show the same level of complexity.

6. Explain why no other species has developed language.

7. Was language much more primitive in the beginning? If yes, then can language become more complex in the future?

8. Chimpanzees, gorillas, and bonobos do not show any signs of having language. Explain why our early ancestors were able to invent language but other primates could not. Were our early ancestors that much different from all the others? How and why?

9. Fossil records show no exponential growth in the advancement of our species up until about 20,000 years ago. Explain how this is possible if language came about gradually over a long period of time.

10. Did other hominid species such as Neanderthals also have language? If so, then how come all of them have gone extinct and why is there no evidence of any high-level abstract or symbolic representations from their species? If not, why is language so exclusive for only Homo sapiens sapiens?

11. Animals of other species do not show any desire to communicate with each other at the level humans do. Explain how our early ancestors—before they had language at any capacity—had this urge to share complex ideas that no other species have shown.

12. Why do we have the inner voice if language is designed for communicating with other people specifically?

13. Is language innate? If it is not, then explain why the critical period exists for language.

14. If language is learned and not acquired, then can we create a language that is fundamentally different from all other languages?

15. The languages that are being used by people living in primitive tribes are just as complex as other languages. Explain how this is possible.

16. Generating and understanding long sentences can become very complex. But how does the human brain process them so perfectly and swiftly?

17. Determine whether our early ancestors were significantly more intelligent than other hominid species before they had language.

18. Are humans significantly more intelligent than apes if language does not exist?

19. Children seem to acquire language much quicker (and much better) than adults do. A child who is not exposed to any language by the time he is 13 years old cannot learn to speak or sign, even if he is not a feral child or was abused by his parents. Did our ancestors invent language while they were still in preteen years or when they were much older?

20. Why do people have trouble with speech when the Broca's area or the Wernicke's area is damaged?

21. Whales and elephants have much larger brains then humans. Does this mean they are more likely to develop language than other species with smaller brains?

22. People born with dwarfism have fairly small bodies and small brains compared to others. But they speak as well as

anyone else. Why does having a small brain does not affect language at all?

23. Neanderthals had larger brains than Homo sapiens. Does that mean they also likely invented language?

24. Did our species evolve gradually to count numbers? For instance, did we evolve to count up to three, then evolve more to count up to five, then evolve some more to count up to ten, and so forth? If not, then why is it different for language?

25. What is the relationship between language, music, and mathematics?

I make an attempt to answer the last question in Chapter 3. I also give some supporting evidence for the Keller hypothesis in the same chapter. (In Chapter 2, we take a look at the human brain and see what it may tell us about language.)

Revisiting Genie

Before we begin the next chapter, let's revisit the question I raised in the preface. Why was Genie, a 13-and-a-half-year-old child from California, not able to learn to speak English? But how come I did (to a certain extent) when I was the exact same age as she was? According to the critical period hypothesis, Genie had to be exposed to language at some capacity, which most likely did not occur. Her father had no desire to raise children and considered them too noisy. Susan Curtiss, a linguist who worked with the feral child, described how Genie remembered her past in her own words:

> Father hit big stick. Father is angry. Father hit Genie big stick. Father take piece wood hit. Cry. Father made me cry. Father is dead.

Suppose she learned to speak English properly, this is what she might have said:

> My father used to hit me with a big stick. He was always so angry. I remember him hitting me with this big stick he had. He took a piece of wood and

hit me with it. I cried all the time. He made me cry so much. But he is dead now.

Brain scans show the brains of neglected children weigh less than the brains of typical children. Neglect often correlates to less stimulation and more trauma. If Genie was neglected and tortured for the first 13 years of her life, it should not be a surprise her brain did not develop fully. Language is one of the areas in the human brain that requires a certain amount of exposure or input for proper development. Unlike Genie's, my brain was fully exposed to language since birth like any other normal child. Even though Korean and English do not have a lot in common, the language regions of my brain were sufficiently stimulated enough for me to pick up a different language, although it took a long time and a lot of effort. (But I still carry an accent and do not speak English as well as other immigrants who moved to the United States before they were 13 years old.)

No matter what the birthplace or the race is, a young child can learn the language of her new environment. Jim Rogers is an American investor who was raised in Alabama. In 2007, he moved to Singapore with his wife. Since then, his daughters—named Happy and Baby Bee—have spent almost all of their childhood in Singapore. Accordingly, they speak perfect Mandarin. Rogers told Forbes Magazine in 2013, "Happy won the (Mandarin-speaking) contest for the whole country and was the youngest in the finals. My little girls grew up speaking two native languages." Happy—a daughter of two Caucasian parents—was so good at speaking Mandarin, she even beat the Chinese children at it. Her father, on the other hand, has not learned the language at all albeit being an intelligent man and a very successful investor.

Having witnessed more than a hundred people in different age groups trying to acquire a second language (L2), I no longer have any doubt language is very much instilled in our brains just like our senses. It is also equally clear that chimpanzees and other animals do not possess the same capability that humans have. We did not choose Nature, Nature chose us. We just happen to be participants in the game. If only Nature had a voice of its own, the Bible could have begun as the following:

> "And Nature said, 'Let there be language' and there was language."

THE BRAIN

"If a man has lost a leg or an eye, he knows he has lost a leg or an eye; but if he has lost a self–himself–he cannot know it, because he is no longer there to know it."

Oliver Sacks

In 2007, Tetsuro Matsuzawa at Kyoto University revealed the results of their experiments on chimpanzees that raised a few eyebrows around the world. Matsuzawa and his research team designed a test to measure chimps' memory potential. The Japanese primatologists trained a chimpanzee named Ayumu to touch numbers on a screen in ascending order. Numbers between one and nine would pop up on the screen in random areas and the chimp would correctly touch the numbers from one to nine without making an error. More remarkably, even when the numbers were replaced with white squares soon after they were up on the screen, the animal continued to perform the task successfully most of the time. When humans took on the same challenge, their performances were not as impressive. They constantly made errors before they could reach five. Matsuzawa and his team of researchers may have showed a glimpse of the truth about our intelligence.

Defining Intelligence

What is intelligence? It may sound like a no-brainer, but no one has yet to come up with a clear definition of the word. Many dictionaries describe *intelligence* as the ability to understand and gain knowledge. For instance, if one were to win a quiz show by answering many questions correctly, he or she would be considered an intelligent individual. But what if a machine beats human contestants? Then should the machine be considered intelligent, perhaps even more than humans?

In 2011, a supercomputer designed by IBM called Watson beat two of the best human contestants on a quiz show called Jeopardy! by a landslide. (The machine's score was higher than the other two scores combined.) In 2016, Google's software called AlphaGo beat Lee Sedol—one of the world's top go players at the time—at playing the Chinese board game. It was not that long ago when it was believed computers will never be able to beat humans at go. This board game is so complex, it has more possible moves at the end of a match than there are atoms in the known Universe. But the machine, once again, prevailed. In 2017, Google announced the retirement of AlphaGo when it became crystal clear that no human can compete against the program.

Even though these machines appear to have such high intelligence, they still do not consciously understand and gain knowledge like we humans do. Machines are still nothing but number-crunching instruments. They are not intelligent like we are (at least for the time being).

The definition of intelligence appears to be more art than science. However, determining the difference between two cognitive powers may be more of an objective endeavor. In other words, we can rightfully declare one to be more intelligent than another entity at a certain cognitive task. But we cannot objectively declare one to be intelligent since we do not know what that actually means. (It is similar to measuring the amount of energy in something even though we do not know what energy is.) A person can boast about her IQ score being 140 to her friend whose IQ score is 90. But it is anyone's guess whether it has anything to do with intelligence at any meaningful capacity. Intelligence is nothing more than one of many human traits such as honesty, physical beauty, and kindness we look for in others. When all the traits come together, we are considered human. But what does it really mean to be a human?

The Sum of One Plus One

A man only known as "Joe" was a patient of epilepsy who had to go through a very invasive procedure to get rid of daily seizures he was having. Corpus callosotomy was performed, physically severing his corpus callosum or the main connection between the two hemispheres. Fortunately, seizures disappeared after his brain split into two halves. The man had become what is known as a split-brain patient.

Michael Gazzaniga is a neuroscientist who has worked with Joe for more than a decade. His research with split-brain patients reveals a lot about our brains. For instance, removing the bridge that connects the right and left hemispheres does not have a detrimental effect on the person's behavior. Even after the procedure, Joe remained the same person with the same personality. He did not lose the ability to speak or the ability to walk. He was—more or less—the exact same person as he was before. But Gazzaniga pointed out a new peculiar behavior Joe now possessed; when he closed his right eye, he could no longer name objects he saw with his left eye.

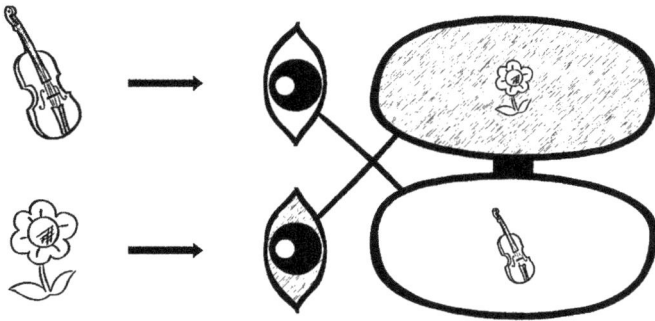

"I see a violin and a flower."

Figure 2-1. One Plus One
The two hemispheres of a normal brain see both objects through the corpus callosum.

Each hemisphere receives information from the opposite side. For example, the right eye sends visual information to the left hemisphere and the left ear sends auditory information to the right hemisphere. But, since the two hemispheres can communicate through the corpus callosum, the brain mostly does not care too much about which side of the body the information came from. However, this was no longer the case for the split-brain patient since the bridge in his brain was surgically removed.

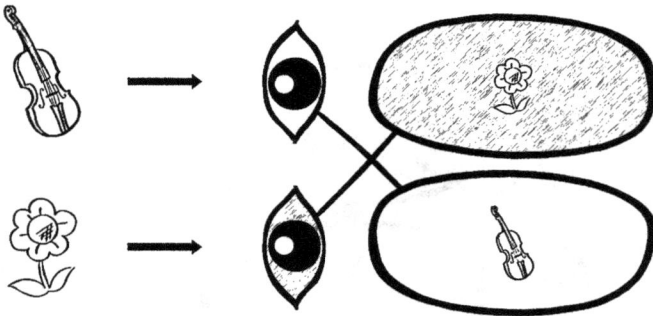

"I see a violin."

Figure 2-2. One and One
The corpus callosum between the two hemispheres is missing. The non-verbal brain (right hemisphere) cannot express what it sees.

So when Joe saw an object with his left eye, the information went to his right hemisphere. But the language center is located on the left hemisphere, which means he could not describe what he saw (even though he clearly saw something and could even draw it). Joe's left hemisphere, therefore, would swear it did not see anything. So which hemisphere was telling the truth? The answer is, well, both.

Joe's right brain saw something with his left eye. That is why he could draw what it saw. But his left brain did not see anything because his right eye was closed. So this part of Joe could not draw what it did not see and Joe would say he did not see anything. (His language department was very likely located in his left hemisphere.) But here is the problem; Joe was still one person and not two different people.

How can one human being have two conflicting realities in his head? Can a person potentially love baseball and hate it at the same time? Split-brain patients like Joe say the answer is an astounding yes. It appears a human can have up to two sets of consciousness in her brain. It turns out two hemispheres can be two different brains. Unlike what the word "hemisphere" might suggest, it is not a half of an organ. In fact, they are "holospheres" that are squished together inside the skull. The two brains operate as one unit merely because they are wired together. We and many other animals actually have two distinct brains each. (This book just gave you one more brain!)

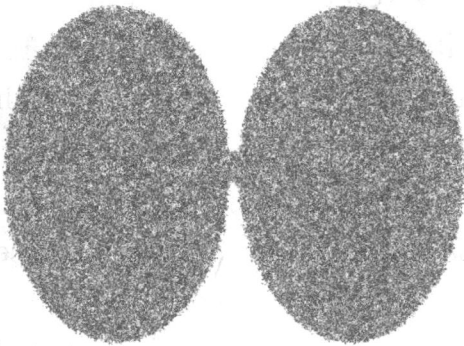

Figure 2-3. One vs. Two
Is it more accurate to describe this image as one unit or two units, especially if it or they can function separately?

The State of Consciousness

Consciousness is another word that is almost impossible to define. We often use this word to mean "awareness of ourselves and our surroundings." As ambiguous as the word may sound, one thing seems to be certain; we would not be able to wonder what consciousness is if we were not conscious in the first place. But what about consciousness of other animals? Generally speaking, carnivores such as lions, bears, and wolves appear to be more conscious than herbivores are. Being conscious helps predators catch prey. Imagine a hungry lion spotting a zeal of zebras not so far away. The lion has two options:

[1] Run toward the zebras immediately
[2] Lower its body and then quietly sneak up on them before launching an attack

According to lions, [2] is clearly a better choice. Zebras are not only fast but also a little too big for lions to catch with ease. This is why big cats usually prefer going after weak, small, or juvenile zebras. But if they react instantly upon seeing prey, they probably would not have been successful enough to survive to this day. It is better for them to be patient and develop a strategy first. In other words, they use their consciousness to maximize their chances of catching food. Without consciousness, it will not be feasible to plan ahead and make better choices.

But not every animal is conscious. It is extremely doubtful insects have the same level of consciousness as humans do. (Some species have nerve nets and do not even have brains.) So it is clear that–as a general rule–the level of consciousness grows as the brain becomes more sophisticated. But if we have two brains per individual, that would mean we could also have two different sets of consciousness. This has never been a concern, however, since the two brains are linked together and the right brain has no voice of its own. So even if the corpus callosum is severed, a person will not generate two different voices inside her head.

There is more evidence that shows distinct sets of consciousness can reside in a person. When one of the brains causes epilepsy, a brain surgeon can remove the problematic hemisphere in its entirety. The procesure is known as hemispherectomy and is known to be a quite successful operation albeit being very invasive. Surprisingly, the patient does not lose anywhere near 50% of his intelligence after the procedure. In fact, most patients show little to no difference in their cognitive abilities, especially since the procedure is usually performed when they are very young. (However, the removal of one brain does paralyze one side of the body. Language also may have to be relearned, depending on which hemisphere is being removed.) Consciousness–like personality and intelligence–shows no significant change for people who have gone through hemispherectomy. It is as if the mind makes an extra copy of all important stuff. But, once again, what is consciousness?

Imagine a small room in a laboratory that has a metal button attached to the floor. Right next to the button is a pipe sticking out from the wall. When the button is pressed, a piece of food comes out of the pipe. But there is a catch in this proposition. The metal button conducts electricity, which will send out an electric shock that is strong enough to be painful but not quite lethal for the animal.

If a simple creature like an earthworm is put into the room, it will not have any clue what is going on. As it roams around the room, it accidentally triggers the button, receiving an electric shock. The worm's reaction is fairly easy to predict. It will move away from the button as quickly as possible. But the worm might return to where the button is sitting, being none the wiser. It is not conscious enough to realize that touching the button causes pain. A rabbit, on the other hand, might be smart enough to figure out the button is responsible for causing the pain. But it may not realize that pressing the button yields pellets. But a cat or a dog would be intelligent enough to know what the jig is. It will consciously accept the pain for a piece of food whenever it is hungry.

Things get a little bit more interesting when a chimpanzee is put into the room. It will not take long for the ape to realize what the setup is. However, unlike a cat or a dog, the chimp can figure out how to beat the system. Using a tool such as a stick or a rock, the primate can avoid getting shocked when pressing the button. In other words, the chimpanzee is not only conscious enough to find the problem, but it can also come up with a solution. But what happens when a human enters the room? Its behavior will be similar to a chimp's, avoiding the negative reinforcement while receiving the positive reinforcement. Taking it one step further, a human–unlike any other animal–might be aware that it is an experiment of some sort.

From an earthworm to a human being, the behaviors observed in the imaginary room might suggest a possibility of consciousness existing in varying degrees. Although it is not nearly as satisfactory as defining the word, it does raise a very important question; can humans have different levels of consciousness? If the answer is yes, then does it explain why some people seem more intelligent than others? Is intelligence simply the result of having a certain degree of consciousness? (So the higher the level of consciousness, the more likely the person will be intelligent.) Also, is there a species somewhere in the Universe that has higher degrees of consciousness than humans?

Figure 2-4. Consciousness Might Be Different for Each Individual
The alpha male, on the far right, is being watched by other stump-tailed macaques. We can only imagine what is going through each one's mind. (Photo taken by Frans de Waal in 1985 at the Wisconsin Primate Center.)

Frans de Waal, a professor of primate behavior at Emory University, believes human behavior is not really all that unique. Waal thinks our behavior is very similar to the behavior of chimpanzees even though we are blessed with language, music, and mathematics. "Language, music, and math have little effect on our day-to-day behavior," said the primatologist. Frankly speaking, our behavior resembles that of other mammals much more than perhaps we would like to admit. The truth might be that language masks so much of what we share with other animals.

Imagine a scenario where the president or prime minister of a powerful nation is deciding whether or not he will launch a nuclear attack against another nation. If he presses the button sitting on his desk, he might be able to claim victory over a preemptive strike. But, as a result, a countless number of innocent people will be killed. If he is not able to be sufficiently conscious about the suffering he can cause, he is more likely to press the button. His advisers might tell him potential repercussions of such an action, but it will not matter if the man in charge is not at a level where he is conscious enough to know what he is doing. He will simply make use of language to make up some justification for wanting to press the big red button. And he just might.

The Recursive Brain

When it comes to intelligence, Homo sapiens and sea sponges are on opposing ends of a spectrum. Whereas humans have taken over the world thanks to our mental acumen, sea sponges really have no achievement. In fact, they do not even have nervous systems. A sponge, which looks like a hollow vase, glues itself in a particular place underneath the ocean. Then it intakes nutrients and oxygen in the water through small pores that does not involve any decision-making process. This automated life style has been very successfully for at least half a billion years.

Sponges may look like neo-impressionistic sea plants. But they feed on organisms like bacteria, which makes them animals and not plants. Interestingly, their genomes show they have genes that are associated with nerve cells. That is why scientists assert sponges once had neurons. If this is true, having brains apparently did not give them any advantage. The fact that they evolved to do without their nervous systems indicates it might be more beneficial for them to not think at all. (In all likelihood, they will outlive our species.)

Creatures like sea sponges show the true nature of evolution. Some species become more complex whereas others refuse to change at all for millions of years. In some cases, they even become more simplistic. This is because complexity does not necessarily mean betterment. It might actually be quite the opposite. Simplicity is a sign of perfection and complexity only arises as a result of imperfection. Species such as tardigrades (water bears) and cockroaches have not changed a great deal in the last hundred million years. They do not possess any special abilities like woodpeckers or hedgehogs do, but they are nearly indestructible.

If one is facing elimination from existence, however, it must change and adapt. Human ancestors had to evolve to move around on two feet when they could no longer survive and reproduce at the same rate as tetrapods. By the same token, the human brain probably did not have to grow so much if we had some superpower like laser beams coming out of our eyes. Our brains became larger over the years to compensate for having such mediocre bodies, at least in theory. The installation of language was either inevitable for our species or it was a gift we randomly acquired by chance. Either way, it made things a lot more complex for our brains. Having the language faculty clearly alters the way we think.

But the process does not seem drastically different from how animals ponder about their own interests. Here is a model of what might be going on in a mammalian brain:

Figure 2-5. Data Processing of the Brain
The brain goes through multiple steps to turn input into output.

Data enters the brain as input (I) through a sensory organ. The brain's memory (M) checks for a match and then the data travels to the consciousness/unconsciousness (C/U) department. Then it goes through data analysis (DA) and the result gets stored in the memory (M). If necessary, the outcome can also turn into output (O). Let's take the following sentence and see how it works:

Je pense, donc je suis.

For animals and many people outside of France, this input probably does not exist in the memory (M). But a human's consciousness/unconsciousness (C/U) recognizes the sentence as a quite terse expression, giving the impression that it sounds like something a French person might say. Since the language faculty–as part of data analysis (DA)–does not process this particular information, the brain will disregard this expression without much thought. But a French-speaking person understands that the sentence means, "I think, therefore I am." Once she makes some meaning of the expression, her memory gets updated. She might even recall that this phrase is credited to René Descartes, a highly-regarded intellectual.

In a sense, thinking can be viewed as a process of data analysis with or without input and output. The core mechanism of our intelligence comes from the memory (M), the consciousness/unconsciousness (C/U), and data analysis (DA).

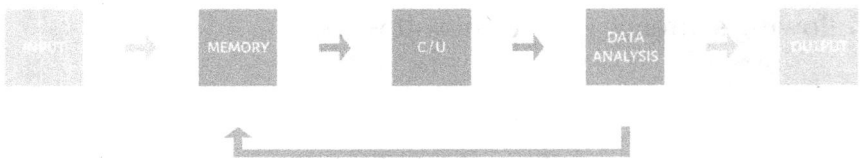

Figure 2-6. Thinking in a Loop
The human brain can create new information from its memory without additional input (e.g. dreaming).

Imagine you are sitting by yourself in a totally silent, dark room. You do not see, hear, smell, taste, or feel anything. In other words, your brains are not receiving any input. But it does not stop you from thinking. You can still recollect events from the past by accessing your memory. As you remember something, you have a chance to add a new meaning to it. Then it goes back to the memory, altering the brain. This recursive process can continue endlessly without requiring any input or output. In fact, it does not appear to be turned off even when we are asleep. Some sort of consciousness/unconsciousness pulls pieces from the memory box and makes up stories as we lie still. Of course, this behavior is not uniquely human. Animals are also known to dream. The difference is our version of data analysis turns abstract thinking into verbal thoughts day and night. Unlike abstract thinking, digital thoughts created by the language faculty can be transferred quite easily between two entities. They can be written down as well, enabling us to increase our collective knowledge as a species over time. What separates us from other creatures is the way our brains analyze the very same world they also get to experience.

A Peek into Our Minds

In 1984 in Arlington, Texas, a writer named Barry Morrow was attending a meeting when he ran into a very unique man with a severe mental disability. This person was so captivating, Morrow later decided to write a screenplay based on him. The script turned into a film called Rain Man (1988). It was a story inspired by a savant named Kim Peek.

Peek was born in 1951 with a mental condition known as macrocephaly, which gave him an enlarged brain. He was missing the corpus callosum and a couple of other connections, completely separating the two hemispheres from communicating with each other. Therefore, Peek was gifted with two discrete brains from birth. As a result of this curse and gift, he became a man who could not forget. Peek had what appeared to be an unlimited capacity of memory. Not only could he memorize all the books he had read, he was also a fast reader. According to his father, Peek was able to read a single page in just eight to ten seconds. It was as if he could scan pages of a book like a machine. But what was more remarkable was the fact that Peek could read two separate pages at the same time. Since each of his eyes was wired to one brain each, the savant could read the right page with his right eye while his left eye was reading the left page. It is unclear whether both of his brains had a language center. But his behavior seems to indicate that, as a matter of fact, they did. Otherwise, it would have been impossible for him to process two different sentences at the same time. But what does that say about his consciousnesss?

Like many split-brain patients, Kim Peek showed signs of having a split personality (no pun intended). He would sometimes throw a tantrum and engage in a conversation with himself as if there were two voices in his head. But this verbal behavior is not common among split-brain patients since most of them have only one brain for language. But the fact that Peek could read books using either his left brain or his right brain may be an indicator that he had two separate voices. The people who knew Peek did not suspect this possibility since he could only say one thing at a time. For him to speak two different thoughts simultaneously, he had to have two different sets of mouths and vocal cords. But he did have two hands he could have used for signing different words. Here is a transcript of Peek speaking to himself (or himselves) in one setting:

Peek: I..I..I do, uh, I..I..I..I..I can't do it.
Peek: (In a slightly different voice and tone) That's what I'm doing!
Peek: Now you're starting, now you know wh...

In theory, Peek could have learned a sign language that requires only one hand in usage. Meanwhile, the other hand would stay perfectly still with the eye and the ear on the same side of his body being blocked of any stimulus. In this fashion, only one of his two brains gets to learn to sign. After training one brain, the other brain can learn it the same way starting from the beginning once again.

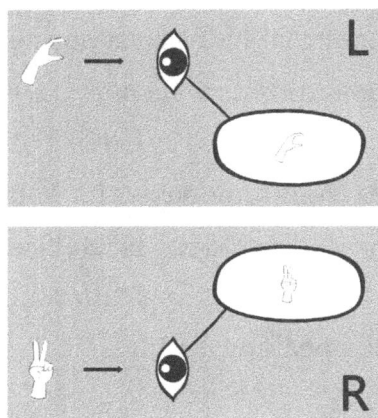

Figure 2-7. One Voice for Each Brain
Teaching a sign language to each hemisphere separately might have given Kim Peek two unique voices.

Once both of Peek's brains are able to talk in sign language using one hand each, we could have observed how Peek's brains would have reacted to each other. It is possible his hands would have signed to each other, making him the first person in history to display communication between two different voices (or two different sets of consciousness) within oneself. This experiment, if it worked, would have shattered the assumptions we have about our minds and the meaning of individuality. (The word *individuality* even ends with "duality.")

Here is another thought experiment that might reveal the truth about our brains and our selves. Imagine two babies, Yin and Yang, being born with unique neurological conditions from two different families. Yin is born with no brain but is healthy otherwise. Yang is born with two normal brains but without a corpus callosum, just like Kim Peek.

Yin's parents are devastated because their child is totally unconscious. He has no mind and perhaps no soul either. Yin's body is well and alive and yet he is as good as dead at the same time. From a distance, Yang's parents see Yin's mother weep in agony. After a long discussion, Yang's parents make a tough decision to donate one of Yang's brains to Yin. The doctor agrees and puts one of Yang's brains into Yin's head, giving Yin meaningful life.

Figure 2-8. A Brain Transplant
After receiving one of Yang's brains, who really is Yin?

Years later, Yin and Yang grow into two healthy boys. Both of them are happy and can talk like any other children. (We assume both brains were capable of properly developing language due to agenesis of the corpus callosum and their young age.) However, one thing remains the same. Yin still has one of Yang's brains and none of his own. Then who really is Yin? Who are Yin's real parents? If Yin and Yang engage in a conversation, is Yang essentially talking to himself? What if the doctor reverses the operation and puts the brain back into Yang from Yin's head? Would Yang become another Kim Peek?

Although this is nothing more than a thought experiment, today's progressing technology might allow brain transplants to take place in the near future, opening a whole new kind of possibilities. For now, we are stuck with a wireless transfer of information, where thoughts of one individual are sent to another person remotely via a medium such as books. (When you are reading this book, your brains are making duplicates of the thoughts that existed in my brains.)

The Computer and the Brain

During the 1930s and the 1940s, three men were developing electronic digital computers for the first time in the world. John Vincent Atanasoff created the Atanasoff-Berry Computer (ABC) with one of his students, Clifford Berry. Then John Presper Eckert and John Mauchly developed several computers such as the ENIAC, which made major improvements to the ABC design. But the most important computer was the EDVAC, also conceptualized by Eckert and Mauchly. Not only was the EDVAC fully electronic and digital, but this cutting-edge machine also made use of a binary system. It was also easily reprogrammable, giving birth to the modern computer architecture.

But there was another man who contributed to the machine's creation. John von Neumann, making him the fourth John, drafted a document explaining the design of the EDVAC. However, von Neumann credited the work only to himself on this historical document, which later became controversial. (The issue easily could have been avoided if he just put down his first name.) Not so surprisingly, the architecture of the EDVAC computer went on to be known as the von Neumann architecture.

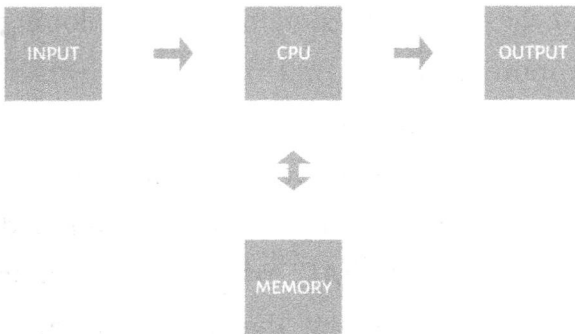

Figure 2-9. The von Neumann Architecture
A simplified version of the modern computer architecture designed by several different Johns.

What is interesting about this schema is that it might give us an idea of how our own brains work. The computer and the brain essentially have the same function; they convert input into output (e.g. 15 + 12 turns into 27). After all, the purpose of the brain is to analyze data and make use of it for our own survival and proliferation. The human brain is essentially an input-output device.

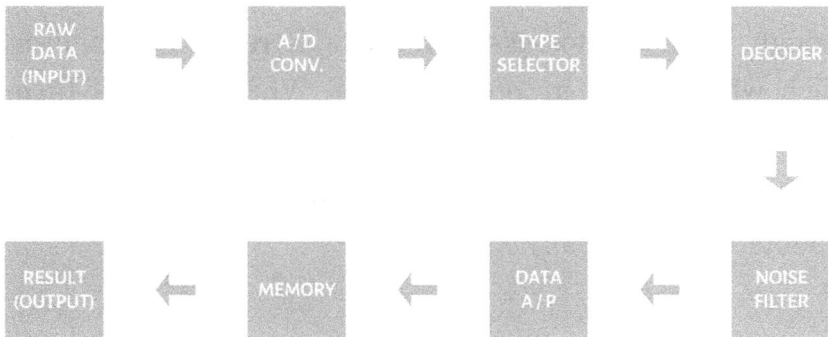

Figure 2-10. The Brain's Architecture?
The human brain, like the computer, processes input data and turns it into output. Here is a depiction of the potential process.

The following is an example of what might be going on inside a human brain. A sensory organ like the eye receives input from the environment. Since the input is analog data, the brain must convert it into digital. Once the data has become digitized, then the brain can figure out its type. For instance, when sound enters the ear, the brain has to determine whether it is a random noise, music, language, or something else. Once the data type has been decided, the data is sent to a decoder so that the brain can understand and analyze the data. But the amount of the raw data might be too expansive for the brain to handle. Therefore, it must first remove nonessential portions of the data by sending it through a noise filter. Now the data is ready for analysis, which will disclose its potential use. When the data turns into useful information, it gets stored in long-term memory for retrieval at a different time. Also, the person can react to it mentally or physically.

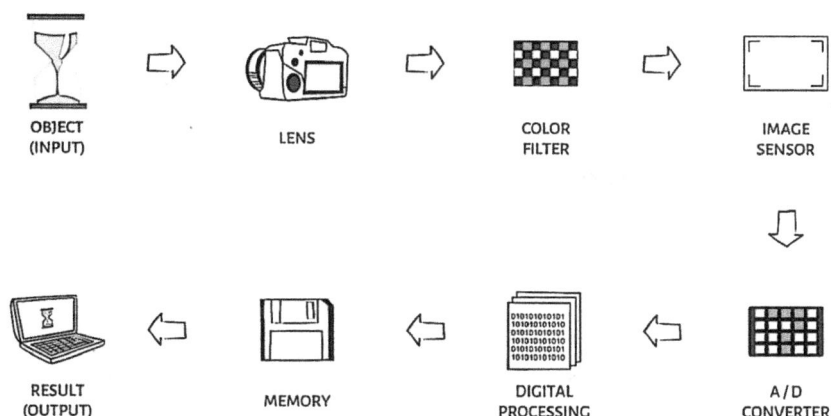

Figure 2-11. Data Processing of a Digital Device
A digital camera takes visual data and turns it into bits of information.

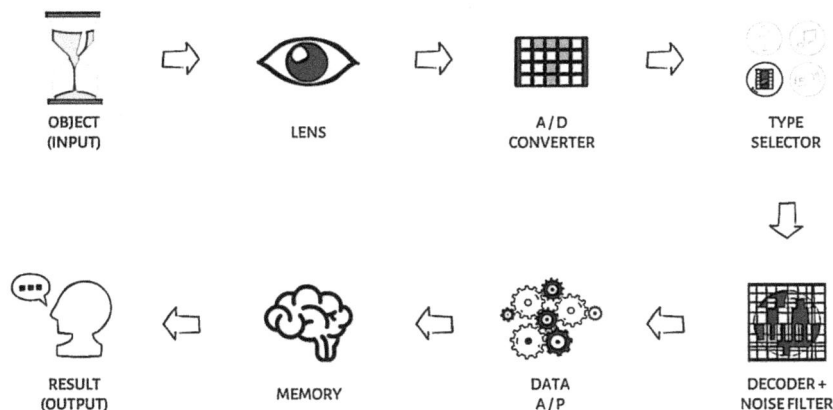

Figure 2-12. Data Processing of the Human Brain
A human brain likely goes through a similar process.

This overall process seems to be more or less the same for most people, except for one part. The process of data analysis varies from person to person and appears to be responsible for producing different types of intelligence. This key step makes every mind unique. In case of Kim Peek, his brains could not analyze data properly. He could read a book and remember every word in it. But he often failed to understand what the book was actually about. On the other hand, people usually do not have any trouble understanding the content of a book but cannot memorize every single sentence. Peek had trouble extracting meaning of other people's thoughts since nothing was getting filtered and everything went straight into his long-term memory. That is likely why he could remember virtually everything, but he could not understand mathematics, social cues, and jokes.

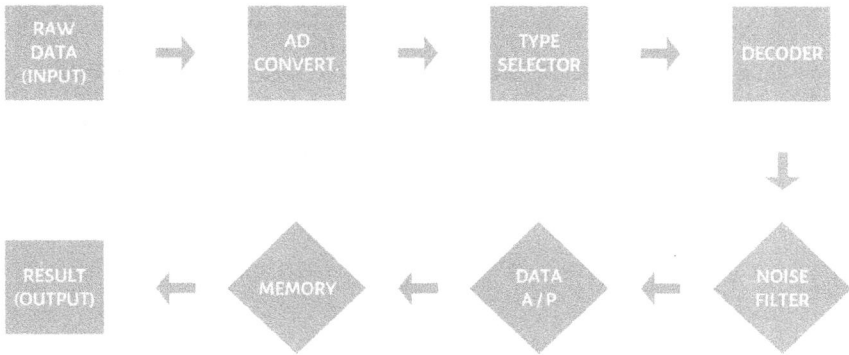

Figure 2-13. A Peek into a Savant's Brain
For Kim Peek's brains, the data processing mechanism appeared to be very unique.

Somewhat similar to Peek, a person in my family was also born with a neurological condition. For him, language comprehension does not exceed the understanding level of a young child. His sensory organs receive data as everyone else, but his brains do not process it the same way. This affects his behavior tremendously, limiting his ability to socialize with people. But speaking has never been an issue for him. Even though he often sounds atypical in terms of what he says, he can still utilize language just like any other human being. Language, it seems, is never half-baked (unless a region in the brain is damaged due to an accident or a stroke). Either one is born with the ability to speak or one does not have it at all. Language looks to be more like a simple bicycle than a spaceship; it requires only a few simple parts for it to work. Apparently, a small tweak in our brains was all we needed.

More than Meets the Eye

Language is rather pure and simple. But that does not necessarily mean nothing else changed in the brain for the sole purpose of language utilization. When people speak, their hands move around unconsciously as if they have a mind of their own. This behavior is observed in every culture and every race, especially during face-to-face interactions. Perhaps the language regions in the brain are hardwired for hand gesturing. Our ancestors might have externalized language in the early days by moving their hands around. Hand gestures do seem a lot more intuitive than spoken words. It is fairly easy to guess what the hand signs might look like for words such as *in, on, to, with, eat, see,* and *run* in any sign language. However, words of an oral language are too random to predict.

	ENGLISH	FRENCH	KOREAN
INTAKE FOOD	"YEET"	"MAHNG-JEH"	"MUHG-DDAH"
PERCEIVE VISUALLY	"SEE"	"VWAH"	"BOH-DAH"
MOVE FAST BY FOOT	"RUHN"	"COO-REAR"	"DDWEE-DAH"

Figure 2-14. The Randomness of Word Sounds
Unlike a hand sign, the sound of a word is usually hard to guess.

Here is one potential scenario of how language was externalized for the first time. A baby, say Wendy, is born with the language faculty due to a genetic mutation. She is the only person with language, but she does not feel she is any different from others. This is because other people also make use of hand signs or spoken words to exchange simple expressions. (It is almost unfathomable to think that our early ancestors did not have any form of communication when many other species of animals do.) So Wendy learns to associate hand signs or spoken words with particular concepts.

Another girl goes through a similar process, but this young child is without sight and hearing. Her name is Helen Keller. Because she was born with language in her brain just like Wendy, Keller was capable of learning to speak or sign. When Anne Sullivan, Keller's teacher, tried to teach Keller a hand sign for the word *water*, the girl understood what words are and how each word has a different meaning. She wrote the following in her autobiography:

> As the cool stream gushed over one hand (Sullivan) spelled into the other the word *water*, first slowly then rapidly. I stood still, my whole attention fixed upon the motions of her fingers. Suddenly, I felt a misty consciousness as of something forgotten, a thrill of returning thought and somehow the mystery of language was revealed to me. I knew then that "w-a-t-e-r" meant the wonderful cool something that was flowing over my hand. That living word awakened my soul, give it light, hope, joy, (and) set it free!

After learning English, Keller went on to inspire a countless number of people. But how did she learn English grammar? How can one individual teach language so successfully to another when the only sensory organ that was used was the skin? Keller could neither see birds nor hear them sing. She could only feel them by touch. And yet, she was able to express how much she loved birds. This feat was only possible because she was exposed to language rather early in life. Wendy, the first human speaker, also had to be exposed to language at an early age. If she grew up with another child who also had the language gene, language development would have been much easier for both of them. Since some hand signs seem to possess intrinsic meanings (e.g. pointing at someone or something), the children could have taught basic prepositions to each other in addition to learning simple nouns and verbs. If they depended on sound as the only modality, language utilization would have been somewhat more difficult.

Whether hand gestures were the primary or secondary modality of language in the beginning, it appears the remnants of our ancestors' behavior still remain today, raising the possibility that early humans without the language faculty still had some form of communication. In general, people use gesturing for emphasizing certain words, especially nouns and verbs. Hand gestures also become more frequent as emotions rise. But this unconscious behavior is not so easy to carry out when one is using a sign language. Since the primary modality of sign language is hand signs, expressing emotions requires another medium. Signers often make very "vocal" facial expressions when they talk, not only to convey meaning but also to show emotions. Otherwise, they would "sound" very cold or lifeless as if they are robots. Conversations are not simply about what is being said but also about *how* it is being said.

Even though hands are very useful for externalizing language, it is nearly impossible to make a claim that our hands have evolved for talking. But the brain seems to have found a way of using them for language. If language was "near perfect" from the beginning as Chomsky asserts, then there would not have been any need for the brain to change anything at all. However, the rest of the brain might have gone through a few modifications to accommodate this very precious gift.

Making Room for Language

Even though externalization of language can take many different forms, speaking is by far the most common one all around the world. It appears the shape of our vocal cords changed over time in order to utilize music, language, or both. Whistling is also a form used for language in different places. People use their hands to sign to each other. Cued speech is a another visual modality where a person takes a spoken language and spells out each word using fingers and mouth movements. We can visually or tactually read written text (e.g. Braille). It seems like any modality can be chosen for language. However, there is one exception. We see little to no use of music as language, even though pitched notes should be able to convey meaning as well as syllables of spoken words can. The reason might be that when the brain picks up melodies, it instinctively selects the data type as music and not language.

Apparently, not all forms are created equal. Verbal expressions are, by nature, time-dependent. Words disappear soon after they are uttered. Most spoken words and phrases are not meant to be stored in long-term memory. Written words, on the other hand, are time-independent. Each letter represents a sound that represents something else in the brain. There is no doubt written language is 100% a human invention and a mighty one at that. But that means reading and writing do not get much "hardware acceleration support" like speaking and listening do. The human brain was never designed for reading books. It explains why people have a knack for learning to speak or sign, but illiteracy is still an on-going problem even in developed countries. Training our brains to decode black line-drawings such as the English alphabet requires a lot of resources, which ultimately may not work properly. Nonetheless, it was the invention of writing that allowed us to keep our thoughts permanently, making books one of the greatest inventions in history of mankind. Without cumulative knowledge being passed down from one generation to another, we still might be walking around barefoot with spears in our hands.

Over thousands of years of evolution, natural selection possibly have given a slight advantage to individuals who were adept at visual symbolic data-processing, enabling us to read without much effort. If this is true, then people who are literate should be considered having a unique skill, which may or may not be a true linguistic feature. An illiterate person could also suffer from reading or writing music.

(Some of the most famous musicians are considered "musically illiterate.") If we had to make room for this talent, something else in the brain might have been sacrificed. The development of a literate brain can very well be different from a brain that has never been exposed to written text. If 'L1' denotes a native language and 'L2' denotes a secondary or foreign language, then perhaps literacy should also be denoted by an additional 0.5. This would put most people in developed countries at L1.5 or higher. Even though language may not be our own creation, we still have enough human ingenuity we can be proud of.

The Early Born Gets to Evolve

It is a mystery why human babies do not develop language earlier in their lives. They develop sight and hearing almost right after birth, but it takes babies about a year before they utter their first words. A few extra years are required for them to express themselves in short sentences. The fact that it takes several years for humans to speak raises a dilemma; if language is innate, then why does it take such a long time for babies to talk? Shouldn't they start speaking by the time they start walking? Of course, just because a baby is not speaking at the 12-month period does not mean his brain is not utilizing language. *Communication* via language takes a couple of years, which is simply a sign that language is manifesting itself. As long as the baby is being properly exposed to language from birth, everything should be fine. But there might be another reason why language externalization gets delayed.

Human infants are notoriously good at expressing themselves. They cry their hearts out and then put on the biggest smile in the world. It is true babies of other animal species are also capable of showing emotions, but they do not come anywhere close to newborn Homo sapiens. The general rule appears to be, the more offsprings depend on their parents, the more expressive they are. It is no secret human babies are completely dependent on their parents for the first several years, creating a need for communication almost right after birth. But they seem to know how to show hunger, discomfort, happiness, and other basic emotions without saying a word. In other words, they already have basic communication way before they develop language.

But when it comes to information processing such as conscious thinking, it is quite a different story. Language becomes handy when infants are old enough to control their own bodies and make decisions on their own. (Picture a little toddler having decision-making power in her brains but not having any control over her body. She will likely end up in a state of delirium.) Most babies start walking when they are around one year old, which is roughly the time period when they begin learning words for the first time.

VOCABULARY SIZE BY AGE (IN MONTHS)

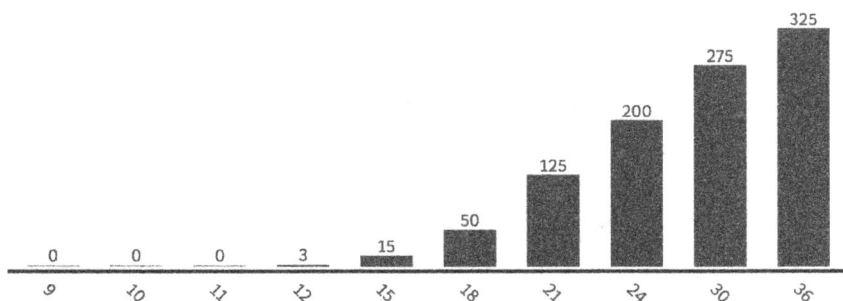

Figure 2-15. Vocabulary Growth of a Newborn
A baby starts learning words starting at around 12 months. By the time she is around 18 months old, she knows 50 words or more. From that point forward, she will learn 50 to 100 words every three months or so.

We also have no recollection of any event that took place before we were three or four years old, a phenomenon known as childhood amnesia. Not coincidentally, perhaps, human brains begin to mature (in terms of size) around three to four years after birth. For chimpanzees, the brain ceases to grow between one and two years of age. The fact that a human brain weighs more than three times that of a chimp brain may give us the reason why it takes more years for our brains to fully develop. The problem is humans are genetically almost identical to chimps. For instance, the gestation periods of humans, chimpanzees, and gorillas are very similar to each other. Birth canals of the three species are also comparable in size (although humans do have the largest).

Unfortunately, because human babies are born with overly-enlarged heads, giving birth has become a lot more painful than Mother Nature had originally intended. Over time, the pelvis of Homos had to change its shape to make the birth canal as big as possible without hampering mobility. So it is possible the birth canal went through a mutation stage to increase the size of the brain. Once the birth canal stopped evolving, Nature had to look elsewhere to continue the enlargement of the human brain; early birth or premature birth. In the distant past, women with the gestation period of nine months likely had an advantage over women with longer gestation periods. The human brain keeps growing rapidly for about two years, making it that much more likely that the mother will not be able to push the baby through her birth canal if the baby is ten months old or older.

Humans are clearly altricial or helpless at birth. However, altriciality is common mostly in small mammals (e.g. mice) or animals that lay eggs (e.g. birds). Big animals like humans are usually precocial and not altricial. Usually, they are ready for the outside world as soon as they come out. This is true even in our close relatives. Six-month-old chimps or gorillas appear to be equivalent to two-year-old humans in terms of their development. Human ancestors might have been precocial like today's non-human primates but perhaps became altricial over time. One obvious reason for this change is the fact that a baby's brain will not be able to pass through the mother's birth canal if he is older than nine months. That may be why it takes almost a year for babies to take their first steps. They also do not start using words until this time period. Hand preference comes even at a later time when the baby is around two years old, which is indicative of when the dominant brain is established. Babies are not supposed to prefer one hand over the other before the 18th month. A baby born with cerebral palsy (CP) will have permanent difficulties with body movement and coordination. One of the most common symptoms of CP is the early development of handedness. Whereas a non-CP baby will likely prefer her right hand by the time she is two years old, a CP baby might show an early hand preference of either hand even before she turns one. This condition may cause one side of the body (and possibly one of the brains) to develop properly but abandon the other side. Cerebral palsy can also cause problems with language development, perhaps due to the early lateralization of the brains. Babies who develop dominant sides at the right time generally display stronger language skills compared to babies who are more ambidextrous.

Even though a human baby typically cannot stay in the womb for more than ten months, it appears that the human brain can use a little extra time nonetheless. Babies do not recognize themselves until they are almost two years old. Suppose a baby is born after being in his mother's womb for 19 months. When he comes out, he should be right on track to walk, learn to speak, and develop a dominant hand. (Whether the mother's body can keep a baby for 19 months is a whole nother story.) This would put us right back into the precocial group with other big mammals.

The Purpose of Language

Many different animal species make use of communication. Even small insects like ants and bees show sophisticated communication skills that have nothing to do with language. Of course, humans are not insects. A human brain has over 300,000 times more brain cells than an ant brain. Since we are a social species like ants are, it is possible language became ubiquitous because we needed a more complex form of communication. By looking at how we use language, we can speculate whether this hypothesis is likely true or not. But, first, let's look at what ants are able to do without having the language faculty and large brains.

Edward O. Wilson, a biologist professor at Harvard University, claims ants are capable of communicating up to 20 expressions to one other. According to the scientist, ants can say things like "Look out!" or "Danger!" by using odorous chemicals. (Researchers also have discovered that some ant species can make squeaky sounds with their abdomens to communicate.) Since humans have over 300,000 times more neurons than ants, a very crude calculation suggests perhaps we should be able to make over six million expressions using the same method. This number seems more than enough for our ancestors to have fulfilled all their needs. For humans of today, we would have to create 200 completely new thoughts every day for our entire lives for us to reach this number.

To be fair, however, human behavior is a lot more complex and not as automated as insects are. Simply having a big number of expressions would not suffice since it would not generate any complex, multi-layered relationships of different concepts. At the same time, simply having a large number of words was probably sufficient for the earliest modern humans. Even though they had fairly big brains, they were still hunting for food like other carnivores. It is far more likely they had a few dozens of simple calls—and not millions of expressions—before language took place. Even modern humans do not seem to utilize over 50,000 words and phrases on average, which is only a small fraction of words listed in a dictionary. (Most dictionaries, abridged or unabridged, contain less than 500,000 word definitions.)

So how do we use language? What is language actually for? Generally speaking, we use language for the following five categories, T.I.M.E.S.:

> Thought: to think, learn, understand, and ask
> questions

Imagination: to predict, plan, remember, tell stories, lie, joke, fantasize, and dream

Music/Mathematics: to sing and use mathematics

Expression: to express one's own thoughts and emotions to oneself or another entity

Social Interaction: to greet, warn, demand, suggest, give orders, and apologize

Putting Music and Mathematics aside, language is split into two distinct groups; (1) Thought and Imagination for thinking to oneself and (2) Expression and Social Interaction for communicating with other people.

According to Chomsky, we use language mostly for thinking. "You can't go a minute without talking to yourself. It takes an incredible act of will not to talk to yourself," said Chomsky during an interview. ("Talking to yourself" is, of course, a form of thinking.) Since we use language to think more than we use it for communication, he argues language must be an internal function. However, this logic seems rather weak. Correlation does not imply causation. We use our hands daily for using electronic devices such as the computer, but they were originally feet that were later modified for grasping tree branches. Just because we use our opposable thumbs to push down buttons on remote controls does not mean that is the natural purpose of the thumb. In theory, language very much can exist for communication and not for processing complex thoughts. In fact, when we are in presence of other people, it is our nature to engage in a conversation instead of thinking quietly by ourselves. Only when we take every piece of evidence into account, we realize language could not have been designed for purposes of communication and social interaction.

Nevertheless, let us look into what we communicate via language a little bit further. When people talk to each other, they are sharing the results of language production. What people voice falls into one of the two types; Expression and Social Interaction. Expression is simply the transfer of information from one person to another, which would not amount to almost anything if language did not exist. On the other hand, Social Interaction is not all that different from what social animals do. Chimpanzees hug each other after having a fight to say they are sorry. Monkeys give out loud warning calls to other monkeys when they sense danger. Ants, bees, and vampire bats are known to share food with hungry individuals within the same colony. It is clear language is not a requirement for Social Interaction.

Expression also would not be so different without language. Babies express their moods in very convincing manners even before they learn to talk. Humans seem to possess the same emotions animals have (maybe with the exception of laughter). That would mean we have always been expressive of our feelings before we became lingual. One thing we could not have done without language is expressing our thoughts. Of course, it is rather unclear whether our earliest ancestors had any thoughts to share in the first place. If they were like other species, the answer would be a firm no. Anyone who has observed chimps and gorillas would admit they do not exhibit a need for high-level thinking like humans. But Genie clearly showed signs of thinking even though she could not speak grammatically. Susan Curtiss, a linguist who has worked with Genie, said the following in a documentary about the feral child:

> (Genie) was extremely interested in everything around her. She wanted to know the word for everything around her. She wanted to engage people all around her. She was not mentally deficient. Her lights were on and everyone who worked with her—from teachers to therapist to me— knew that she was not retarded. It was clear as day.

Even though Genie never developed language as a young child, she still had to have some language-related areas left intact in her system. The only problem was that some, if not most, of the functionality faded away due to lacking exposure to language before she became a teenager. Nevertheless, no chimpanzee or gorilla has ever shown the same kind of acuity and curiosity as Genie. If language came from a mutation, then the mutation likely gave us discrete thoughts. Externalization of language such as speaking only discloses them. In a way, it is somewhat similar to music and deafness. Even though deaf people cannot hear, they still seem to be attracted to the *feel* of music. Many deaf people enjoy going to concerts to feel haptic vibrations generated by loudspeakers. (They achieve this by being barefoot on the ground or holding half-full water bottles.) Some even become drummers or percussionists even though they cannot hear any sound.

So if Thought and Imagination are internal uses of language and Expression and Social Interaction are not directly related to language, then it means only one thing; communication and language are not as dependent to each other as they may seem. What language gives us is a higher degree of imagination and thought. Stuck on the surface of Earth, we can still imagine what our planet might look like from outer space. But a chimpanzee looking at a photograph of Earth will never realize what he is seeing, not that he is curious by any chance. He and other chimps belong with all other animals while we march on by ourselves.

Music and Absolute Pitch

Language aside, humans also have a common cognitive aptitude in music. Many people have attempted to find similarities between language and music, hoping that these two mental faculties are essentially the same in some way. Neuroscientists have found some common neural activity grounds in the human brain, further enticing the notion that music is language and language is music. Unfortunately, empirical evidence suggests that this is likely not the case. A patient suffering from amusia—the inability to comprehend or appreciate music—does not have any trouble with speech. Another person may not be able to speak but can sing just fine. Kyle Coleman, in the United Kingdom, was born with autism and could not learn to speak. (He was diagnosed as non-verbal autistic and could only say a few words in some occasions.) But a revelation took place when he was around 25. One day, Coleman's mother realized that her son can sing and sing lyrics. Music enabled him to utter words out loud just like everyone else.

How the human species was gifted with two extraordinary cognitive features is a big mystery and the answer may never be discovered. It is believed that our ancestors had music before they had language, but nothing has been proven so far. It does seem feasible that something like music had to be an antecedent to language. Otherwise, language could have happened to multiple different species. But, in reality, language belongs to only one group of animals when there are millions of other potential candidates. It is true that some birds can sing, giving the appearance that they might also develop language one day. But when birds "sing," they are essentially shouting mating calls on repeat mode. It is the same speech over and over again. Their sound utilization is way too simple to be considered as either music or language.

One of many interesting aspects of music is absolute pitch. Also known as perfect pitch, it is the ability to recognize exact notes instantly upon hearing them. For instance, a person with absolute pitch can listen to a melody and recite the notes as if she is reading them off a music sheet. This peculiar musical talent can be applied analogously to language. Even though a parrot can repeat a short phrase spoken to him, it still does not process the words in the phrase the same way a human does. A human being not only recognizes the words but also understands the meaning of the whole phrase. Comparably, while people without absolute pitch can repeat a tune they just heard, they are not consciously aware of *which* notes or chords they heard. The only thing they can do is sing back the tune just like a parrot. But people who have acquired absolute pitch know exactly what they are listening to. For them, music is truly another voice in their brains. For the rest of us, we only have absolute speech. Of course, one does not need absolute pitch to enjoy music. What is interesting is a critical period seems to exist for absolute pitch the same way a critical period exists for language. For someone to acquire absolute pitch, he has to be exposed to musical notes and chords consciously while he is very youthful. If we can uncover the mysteries of music, we might be able to finally discover what makes the human brain so unique.

Language's Side Effect

Millions and millions of years ago, all hominins were happy little apes. None of them could speak, but they were not lacking all cognitive aptitudes. Like other predatory mammals, they had analytics. It might sound a little strange, but many species of animals possess a natural talent to analyze visual data on the fly. For instance, a cheetah uses analytical skills to pick a gazelle from the herd, gets near it without being discovered, starts chasing it, and catches the prey. Since gazelle are almost as fast as cheetahs, big cats must depend on analytics to avoid starvation. The way chimpanzees hunt small monkeys proves they are also very analytical. It is hard to imagine our ancestors being the only predators with no analytical skills, especially considering the fact that their lackluster bodies were almost useless for hunting.

Figure 2-16. Analytics at Work
When a baseball is moving in the sky, one has to use his non-verbal analytical skills to predict where it will land. People with very poor math skills tend to have difficulty catching a ball in movement.

Today, analytics is used by stockbrokers, investors, doctors, and researchers. Professional sports teams also make use of analytics as part of their winning strategies. During a game, the athletes become analytical when they make real-time decisions such as kicking the ball toward the goal. Analytics is very useful but does not require language to operate, which is why animals can utilize it so well. Without analytics, an eagle will not be able to catch live squirrels. It almost seems as if analytics is a primal form of mathematics. Once language came along, only then we were able to invent (or discover) math by working analytics and language together.

Unlike analytics, however, mathematics is not all that fun. Not only is math confusing for many people, but it is also too theoretical and unimportant for us to actually care about it. There is simply not much to like about math. It does not seem to invoke any natural human emotion except perhaps boredom. Nonetheless, mathematics is a quite remarkable invention (or discovery). Without it, we would not have a plethora of useful things such as modern airplanes, lights, and smartphones. One can even argue it is predominantly math, and not language, that gave us such advanced technology. If we have never had any mathematical skills, we still might be hunter-gatherers. However, thanks to language, we are able to harness the power of mathematics. Whatever the cause was for the insertion of language in the human brain, it was clearly a wondrous gift for our species. Therefore, we should not discourage ourselves when it comes to finding the origin of language, no matter how difficult the objective might appear. Until we know what it is, we may never fully understand what we really are.

THE NEUROCOGNITION

"The whole is other than the sum of
the parts."

Kurt Koffka

So what is language? According to many dictionaries and encyclopediae, the word *language* is defined with words like *communication, communicate, community, express,* and *expression.* At the same time, words such as *think, thinking,* and *thought* are usually missing all together. The consensus tells us most people view language as a method of communication for interacting with other humans. It almost seems like a no-brainer; obviously, language has to do with communication. What else can language be about? But just because something quacks like a duck does not necessarily mean it is a duck.

Here is a thought experiment that might give us a better idea of language's true definition. Let's say a woman named

Selena invented a machine that can wire two people's brains together. Friends of Selena, Lenny and Ricky, volunteer to be the guinea pigs for this experiment. All three of them believe the machine will connect Lenny and Ricky's brains in a way that will allow the two men to communicate with each other directly. They do not have to open their mouths to talk to each other. They can just sit tight and let their brains do the talking. But it turns out that is not what actually happened. Once the machine started running, Lenny and Ricky did not hear each other's voices. Instead, they had two independent thoughts occurring at the same time. But neither Lenny nor Ricky could identify which one is his and which one is not.

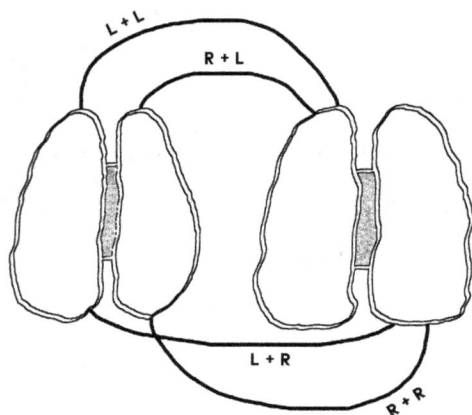

Figure 3-1. Wiring Two People's Brains
Connecting four brains requires four separate connections.

When two people's brains are connected directly, they bypass any medium such as sound or light. That means no communication takes place between them. No words have to be spoken or signed. Communication occurs only when two different entities are connected through a medium. Without a medium, there is no data sent or received. Therefore, the brains of two different people can communicate with each other only when they are separated. In a similar fashion, two hemispheres of an individual are linked by the corpus callosum, but communication does not take place since a direct connection is present.

So when the brains of Lenny and Ricky are linked together, the men become one conscious being, Lenny-Ricky. But, since Lenny-Ricky has two language centers, he would either have two different thoughts simultaneously or every thought would occur twice. (If there is a delay in data transfer between the two language centers, Lenny-Ricky might hear an "echo" every time he has a thought.) This means language is a mechanism that creates sentences or thoughts by processing data generated in the brain. Unfortunately, linguists and neuroscientists have yet to figure out the inner workings of language in the brain. That gives us an opportunity to see it for ourselves.

The Mystery of Word Order

For many years, I wished English was a lot more like the Japanese language. Then learning it would have been much easier for me, whose native language is Korean. Unlike English, Korean and Japanese share the same word order. The main word order of English is SVO, Subject-Verb-Object. Here is an example:

Molly sings the blues.

This sentence starts with the subject, *Molly*, which is followed by the main verb, *sing*. An object noun, *blues*, comes last. Languages such as English, French, Russian, and Spanish all fall under SVO. Languages such as Korean and Japanese are considered SOV, Subject-Object-Verb. Here is the same sentence in SOV:

Molly the blues sings.

Out of total six combinations (SOV, SVO, VSO, VOS, OVS, and OSV), most natural languages belong to SVO and SOV. A few languages like Hawaiian have the word order of VSO. The remainders are VOS, OSV, and OVS, which are even more uncommon. (It is not exactly clear why only two of the six word orders are overwhelmingly popular. Perhaps it is due to the fact that the subject is the main focus of a sentence, making SVO and SOV seem more natural.) The question is, why do different word orders exist in the first place?

There is no doubt humans are all equipped with the same ability when it comes to language acquisition. An Indian born and raised in Wales would speak perfect Welsh. A Belarusian who has spent all of her life in Mongolia would speak Mongolian just like anyone else in the country. Humans are capable of learning any language as long as they start out early in their lives. At the same time, languages appear to be quite different from each another. Here is an example of a Korean sentence:

Bab meokgo ga.
("Bahb muhg-ggoh gah.")

'Bab' means *meal* in English and 'ga' means *go*. 'Meokgo' is a combination of two words, *eat* (meokda) and *and* (-go). So the sentence in English—in the same order—looks like the following:

Meal eat and go.

In Korean, the subject of a casual sentence is often omitted and implied. Also, articles such as *a* and *the* are not required and plural marking of nouns is not necessary. So "Meal eat and go" can be translated as "Have a meal before you go" in the SVO order.

Because Korean is SOV, the main verb at the end of a sentence often gets combined with the suffix '-go' (it means *and*), which is then followed by another verb. But combining multiple verbs with the word *and* is not as common in English. Clearly, having different word orders results in having different types of rules. If this were not the case, learning foreign languages would not have been such an impossible task. But there is a silver lining. Most languages fall under two types of word order, SVO and SOV. It turns out these two may actually be identical underneath the surface.

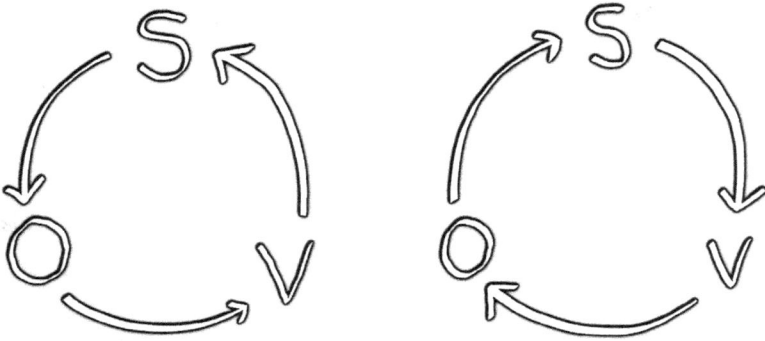

Figure 3-2. One or the Other

All natural languages seem to travel in one of the two possible directions (clockwise and counterclockwise).

SVO languages like English move in one direction, say clockwise, whereas SOV languages like Korean move in the opposite direction, say counterclockwise. Here is one example:

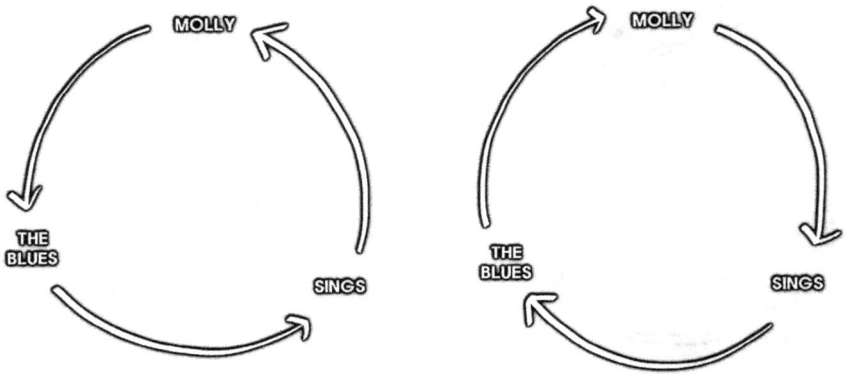

Figure 3-3. A Simple Sentence in Two Directions
SOV languages and SVO languages flow in two opposing directions.

But this schema completely breaks down when a sentence becomes more complicated. Sentences with more than a few words such as "Molly sings the blues with her heart" do not seem to have two working directions of flow.

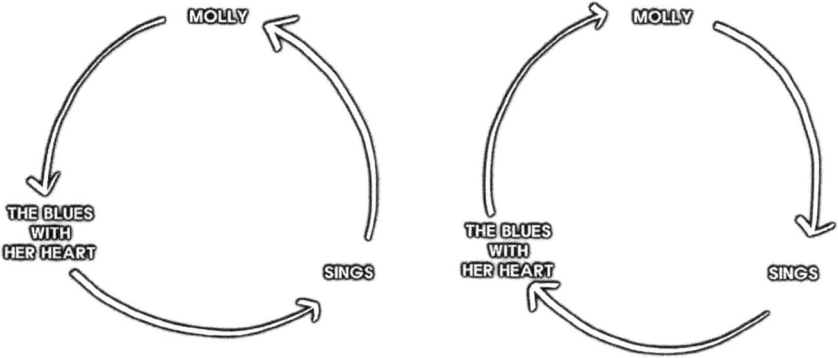

Figure 3-4. Complexity Creates Problems

A more complex sentence no longer works in this fashion.

If this sentence is translated into Korean, the correct word order should be the following:

Molly her heart with (the) blues sings.

Thus, it appears SVO and SOV languages are not two sides of the same coin. This also indicates that the way the brain processes English sentences is quite different from the way it processes Korean sentences. It is almost too obvious; different languages have different rules of grammar. But taking a closer look at it actually shows the brain might be handling English and Korean the exact same way, suggesting that the two languages use the same process. In other words, all human languages might be essentially the same.

The Synapper

Before we can unify all human languages in the world, we must first figure out the process the brain uses for handling language data, which I call the synapper. The synapper is responsible for processing a string of words into sentences or verbal thoughts. Since no one knows what the synapper looks like, our objective is to come up with a candidate. Because language data is carried out by neurons, the synapper likely resembles the way brain cells are connected to each other. Here is a depiction of the synapper:

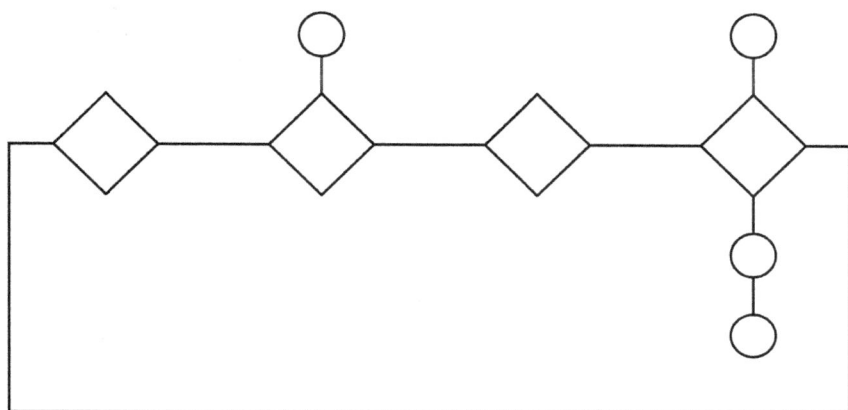

Figure 3-5. A Synapper Model
Like a neural network, the synapper is a closed circuit where information travels in one way or the other.

This modeling technique of the synapper shows data traveling in one of the two possible directions like a clock. The stem or main connection has nodes that are linked left and right (diamond shapes). A node can have branches in a different dimension such as up and down (circle shapes).

Typically, it is not easy to translate sentences such as "Molly sings the blues with her heart" from English (SVO) to Korean (SOV). But something happens when the sentence is depicted in two dimensions.

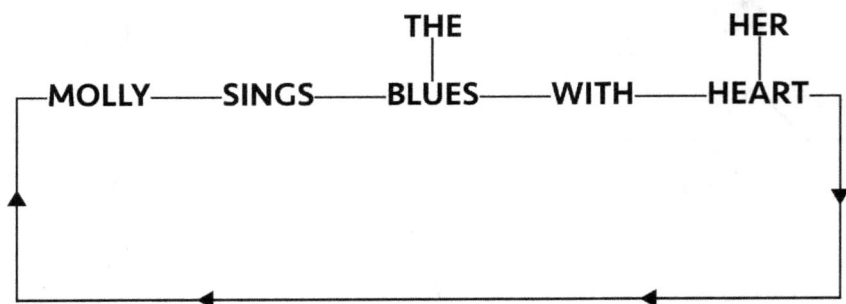

Figure 3-6. Language in Two Dimensions
The words *the* and *her* are connected to their "parent words" (nodes).

Because *the* in the sentence is associated with *blues* and *her* is associated with *heart*, we can move them in another direction. Now these two modifiers are only linked to *blues* and *heart*, respectively. Since English (SVO) and Korean (SOV) appear to operate in opposite directions, the direction of flow should change from clockwise to counterclockwise.

If the sentence is read in a counterclockwise manner starting from the subject, then we have the following:

Molly her heart with (the) blues sings.

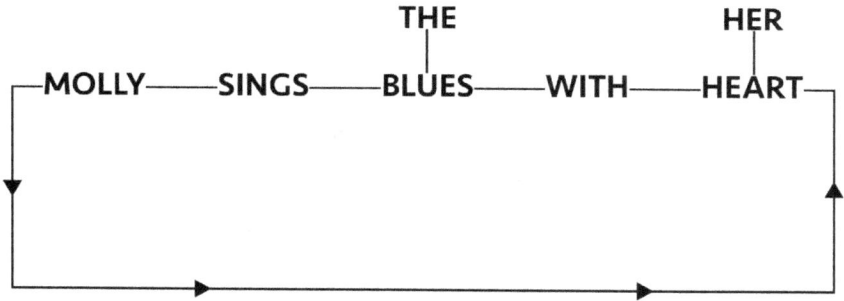

Figure 3-7. English to Korean in a Single Step
By switching the direction, the English sentence now has become a Korean sentence.

This way, just changing each word into Korean results in a perfect translation of the sentence. (The word *the* can be removed since articles are not necessary in Korean.) By taking language into a higher dimension, the sentence is now unified into one form for both English and Korean, two languages once considered nothing alike. The only difference is the direction each language takes. Here is another example:

A curious mind is never bored.

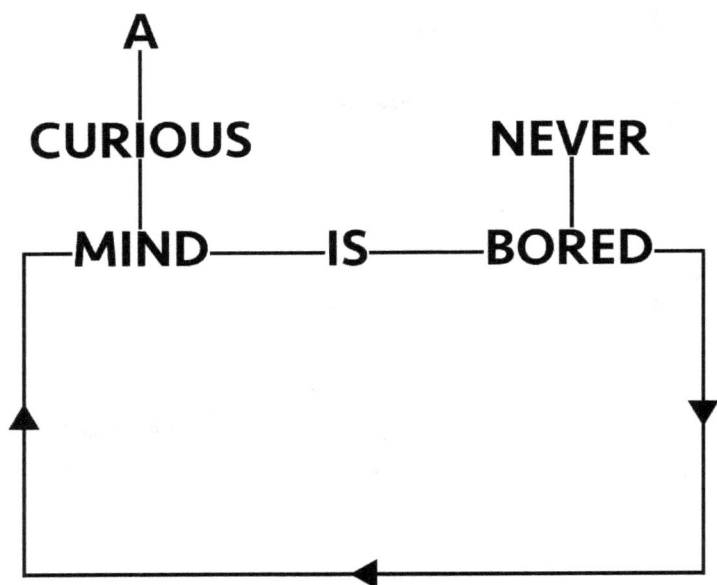

Figure 3-8. Clockwise for English (SVO)
Modeling of the synapper works with any sentence.

To convert the sentence from English to Korean, all we have to do is change each word into Korean and read the sentence in the opposite direction (starting with the subject).

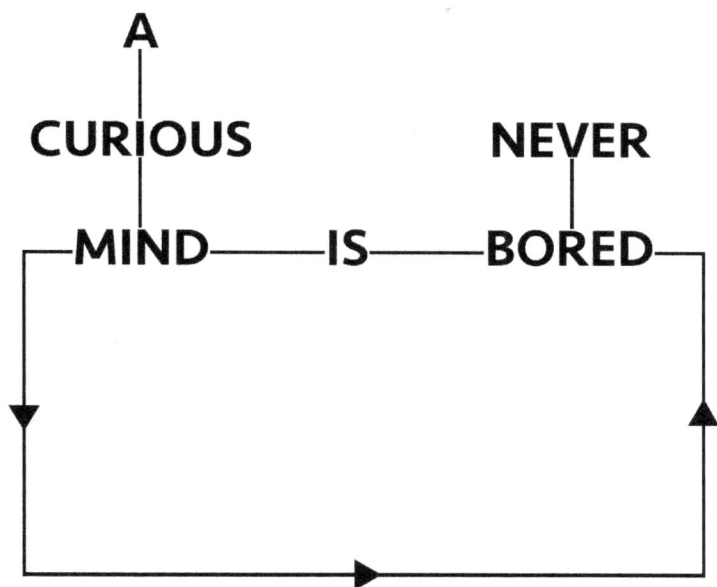

Figure 3-9. Counterclockwise for Korean (SOV)
The correct word order and translation in Korean can be produced by switching the direction of flow.

The article *a* can be omitted, but it is left here to show where it should be. The Korean translation of the sentence becomes the following:

(A) curious mind never bored is.

Once the words are replaced with Korean words, then the translation basically becomes complete. Here is another representation of the sentence:

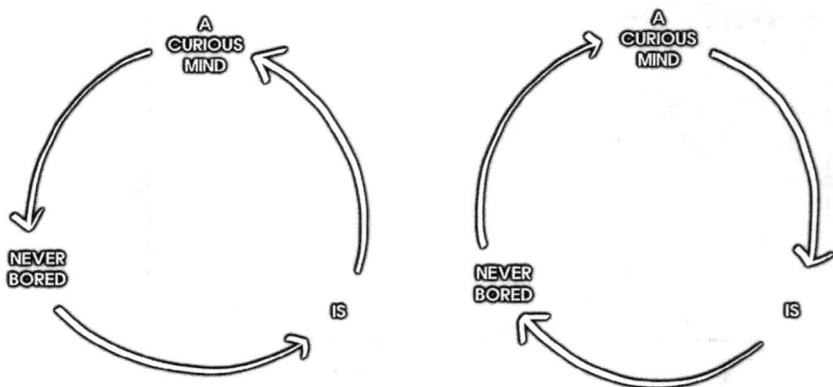

Figure 3-10. Language Processing in the Brain
Is it possible all human languages go through the same process in the brain?

Whether this modeling method of the synapper works for every language remains to be seen. At the same time, the fact that sentences in two vastly different languages can be unified into one for each syntax-semantics representation is a compelling indicator that Chomsky's theory of universal grammar (UG) is in fact true. Language is indeed an innate trait of our cognition. But "languages" are unmistakably human inventions. It was humans who decided that *are* is the right form for the second-person pronoun *you* but not for singular third-person pronouns like *he* and *she*. Because of many artificial rules we have come up with, different languages have different rules of grammar even if they all belong in the same word order. In French, for instance, modifiers can be placed either before or after the noun unlike in English. Both *un curieux esprit* (a curious mind) and *un esprit curieux* (a mind curious) are proper phrases. (There are also English sentences such as "The food here is really good," where the modifier *here* comes after the noun.) This is why children's speech is often "corrected" by people who have graduated from grammar school. Ironically, it could very well be young children who are the ones utilizing language most accurately. Whatever the case may be, synapper modeling does hint the possibility of all human languages using the same underlining mechanism inside the brain. It even works with recursion. Here is an example of a recursive sentence:

I know this sentence is recursive.

"This sentence is recursive" is not recursive. But, by adding "I know" in the beginning, we now have a recursive sentence. This sentence can become even more recursive by adding phrases like "it is true" or "he said." (e.g. "I know he said that it is true this sentence is recursive.") Linear translation with synapper modeling no longer works when recursion is present. But the problem swiftly disappears when recursion is applied in layers.

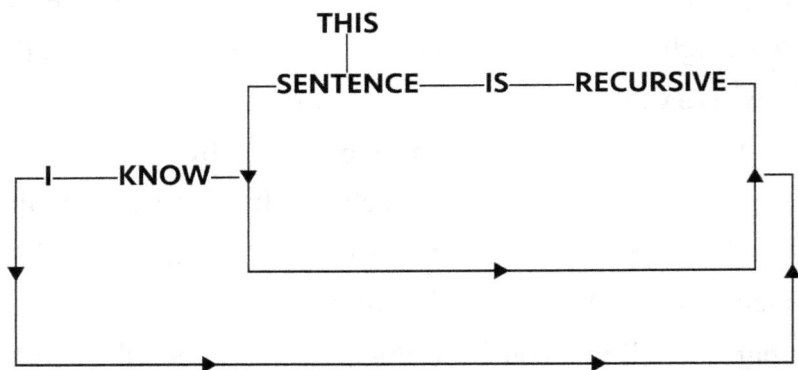

Figure 3-11. Recursion in Synapper Modeling
Synapper modeling is recursion-friendly.

The correct word order of this sentence in Korean is the following:

I this sentence recursive is know.

Starting with the subject, *I*, the direction of flow then tackles the inner sentence as a whole before finishing with the main verb, *know*. Synapper modeling gets the right word order perfectly even when the sentence is recursive. If it does turn out this method of language processing works for virtually all human languages, then it might be the undeniable proof that language is innate after all.

Numbers in Another Dimension

Modeling of the synapper appears to be quite effective in two dimensions. Perhaps the supposed genetic mutation did not alter the brain significantly except for enabling words to be connected in a slightly more complicated way. Animals that do not possess this attribute are not able to have language for two apparent reasons. First, their brains are likely only capable of forming one-dimensional neural circuits unlike ours. (Neanderthals also might have had some "language" in 1D.) Second, even though animals' conscious minds might be able to acquire language in theory, it would be an extremely difficult task without having a native language. A human child is capable of learning a second or foreign language (L2) only if her first language (L1) is intact. But if the child is well into her teenage years, her brains most likely will not pick up a new language; therefore, she can only construct sentences using her consciousness. Genie clearly showed signs of human intelligence when she spoke. But her speech did not showcase proper English grammar; she could only put words together in what appeared to be a linear formation of sentence structure. In other words, the synapper did not get to develop properly in Genie's brains.

Again, the synapper is defined as the mechanism of the human brain responsible for generating syntax. On the other hand, modeling of the synapper shown in this book is purely a speculative demonstration of what the synapper might look like. Using this modeling technique, the synapper is presumed to have a main connection of core words called *nodes* that are strung together in a particular direction. *Branches* are words like modifiers that are connected to nodes in one or more opposing directions.

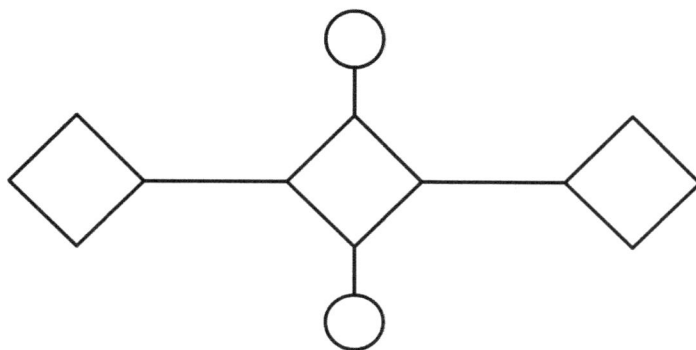

Figure 3-12. Synapper Modeling in 2D
A synapper model in two dimensions showing three nodes (diamonds) and two branches (circles).

But not all modifiers are created equal. For one, modifiers can be split into two groups; words and numbers. Since words like adjectives add qualities to nouns, we shall call them *qualifiers*. Numbers are *quantifiers* since they quantify noun objects. (Words like *few* and *many* do not represent specific numbers, so they will be considered as qualifiers in this book.) Here are examples of the two groups:

Qualifiers: fast, red, the, many, half, very, all
Quantifiers: 12, zero, fifth, 273, once, 4/9, π

By separating numbers from other modifiers, we need to use another direction specifically for numbers. This means synapper modeling now requires three dimensions.

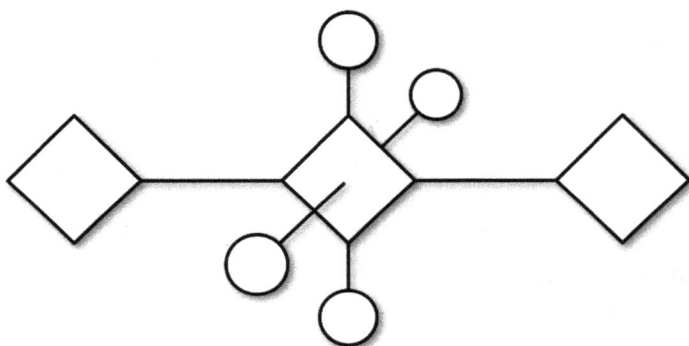

Figure 3-13. Synapper Modeling in 3D
A synapper model in three dimensions. Nodes are in the left-and-right direction. Qualitative branches (qualifiers) are in the up-and-down direction. Quantitative branches (quantifiers or numbers) are in the back-and-forth direction.

Separating modifiers into two groups may seem like an unnecessary step. But numbers seem unique enough, differentiating them from other words may be the correct move. (However, numbers will be treated the same as other modifiers for the remainder of this book for simplification.)

Our ability to understand numbers likely far precedes the birth of language. Even animals have shown an innate sense of numbers, commonly known as number sense. Whereas most species can count up to four, Homo sapiens are capable of counting to infinity due to having language. But even we humans do not sense large numbers intuitively except for a few savants. Most people see numbers between five and ten without counting them one by one.

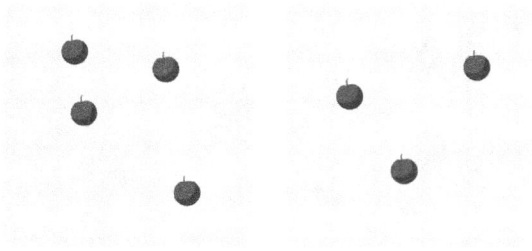

Figure 3-14a. Number Sense Test (1/4)
When numbers are fairly small, it is quite easy to tell which side has more units (relative number sense) and how many units are on each side exactly (absolute number sense).

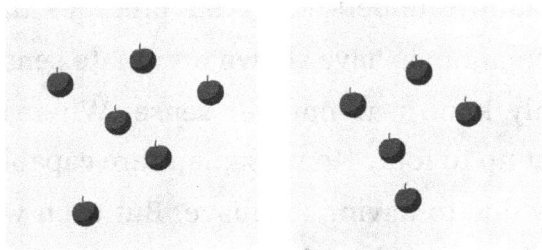

Figure 3-14b. Number Sense Test (2/4)
As numbers get larger, it becomes more and more difficult to sense them intuitively. However, most people are still able to guess which side has more units.

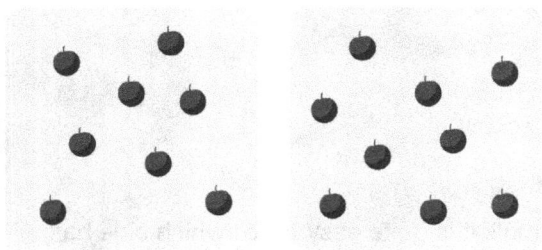

Figure 3-14c. Number Sense Test (3/4)
When numbers get close to ten or larger, many people's absolute and relative number senses begin to err.

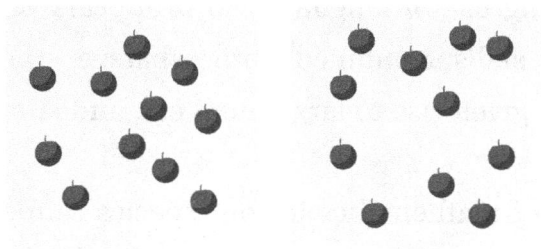

Figure 3-14d. Number Sense Test (4/4)

Some savants can "see" large numbers instantly without having to count.

Absolute number sense is the ability to recognize a number instinctively. Relative number sense is the ability to perceive which side has more units. This analytical talent is independent of the language faculty (at least in animals), which means humans are not likely to outperform other species by a wide margin at this particular task. But a few lucky ones are born with much superior number skills. Oliver Sacks, a British neurologist, shared his experiences with people who could see numbers such as 79, 111, and 183 without counting. These savants are born with a natural ability to see large numbers intuitively. Whether this gift has anything to do with language or the synapper is not certain. Then again, developing the concept of numbers would not be feasible without making use of language. So it appears our primordial analytical skills combined with whatever that gave us language have given rise to large numbers, and later, mathematics.

The importance of mathematics for our species cannot be overstated. It is not just having words but also having numbers that allowed us to harness such advanced technology. If it were not for numbers and math, even something relatively rudimentary like farming might have been too difficult. Interestingly, humans seem to possess different levels of aptitude for mathematics when virtually everyone has language at the same level. This might suggest processing mathematics in the brain is rather complex and not as simple as it is for language.

Patients of brain disorders show the brain may be handling number words in a distinct manner compared to other types of words. People who suffer from Broca's aphasia tend to have trouble speaking or signing. In one instance, a patient suffering from Broca's aphasia could only say one word, a condition known as monophasia. For him, the word was "toto." When someone asked him a question, he would answer it by saying something like the following:

Toto toto toto, toto toto toto toto toto.

In his mind, he perceives his speech as perfectly normal. However, for some peculiar reason, he could still count numbers out loud normally. When someone asked him to count, he started counting with his fingers and said, "One, two, three, four, five, six..." with clarity. After eleven, he became confused and slowly went back to repeating "toto." Another patient developed a quite different language impairment after having a stroke. Even though she could get herself to talk cohesively (with a bit of a struggle), saying numbers was another matter. Even after receiving years of speech therapy, her ability to speak numbers did not show much improvement.

In some languages like Korean, numbers or numerals are considered a separate part of speech. These words might show the appearance of typical nouns, but the brain may require more resources for processing them as part of language. Even professional news anchors sweat from time to time when they have to read large numbers. The synapper evidently seems to have some impact on our primitive analytical skills, potentially giving birth to mathematics. If math can be affected by the presence of the synapper, then it also might be case for our comprehension of music.

Show Me the Music

Three distinct cognitive functions of the human brain–language, music, and mathematics–set us apart from other species. Currently, little is known about their neurocognitive infrastructures as further advancement of technology is required in neuroscience to truly unveil how they work. From what we know so far, it does seem like language and music share many similar features. Thus, we should perhaps put them to the test to see whether they actually have the same underlying mechanism such as the synapper. But let us first compare properties of language and music:

> Element: syllable / note
> Base unit: word / motif
> Group of units: sentence / melody
> Syntax: present / present
> Recursiveness: present / present
> Reversibility: present / present
> Meaningfulness: present / present
> Symbolism: present / present

Words are made of syllables the same way motifs are made of individual notes. A group of words makes up a sentence and a group of motifs makes up a melody. Both sentences and melodies have syntax. We can tell when a sentence is grammatically correct or not. Similarly, we can also detect a wrong note being played in a melody. (I should mention that some musicians such as Thelonious Monk do not believe wrong notes exist.) Both language and music can have recursion, too. An SVO sentence can be reversed into an SOV sentence. Playing music in reverse still has the impression of music (although it often sounds psychedelic). But does music show meaning or symbolism like language does? If we believe words like "I love you" have meaning, then singing a serenade to a love interest could be considered meaningful as well. A string of notes can also be symbolic. For instance, hearing as few as four notes of Felix Mendelssohn's Wedding March or Richard Wagner's Bridal Chorus makes us think of a wedding. If language and music are indeed very much alike, then perhaps the synapper can be applicable for music as well.

If music is anything like language, perhaps notes can be divided into nodes and branches. A node with its branches would become a motif, a short melodic piece. Combining sets of nodes and branches could give us a melody like the following:

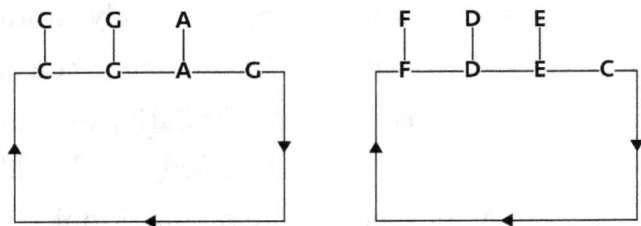

Figure 3-15. The Synapper and Music
Synapper modeling showing the first two melodies of Twinkle Twinkle Little Star.

The first 14 notes of Twinkle Twinkle Little Star are divided into two melodies or eight motifs:

C-C G-G A-A G
F-F E-E D-D C

Playing the nodes without the branches still gives us the same feeling of the song:

C   G   A   G
Twinkle twinkle little star

F   E   D   C
How-I wonder what-you are

Here are the first 23 notes of another children's song, Itsy Bitsy Spider:

D-G-G G-A B B B-A-G A-B G

B B-C D D C-B C-D B

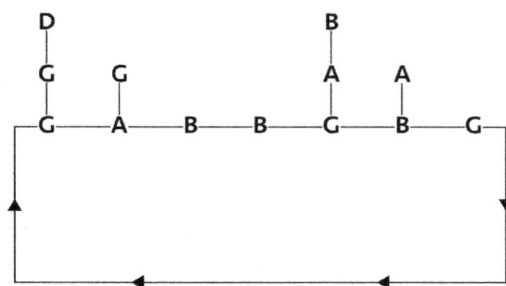

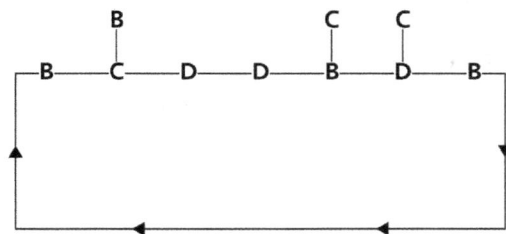

Figure 3-16. Itsy Bitsy Spider

The first half of Itsy Bitsy Spider shown in two synapper models.

This time, things are more complicated because the branches are different from the nodes. But playing the nodes without the branches still gives off the same mood of the song.

G   A   B B   G   B   G

The-itsy bitsy spi der went-up-the water spout

B   C   D   D   B   D   B

Down came-the rain and washed-the spider out

Interestingly, switching the progression of the original 23 notes in the opposite direction—the same way an SVO sentence turns into an SOV sentence—no longer displays the same musical appeal. It still sounds very much like music, but it does not sustain the pleasing quality of the song being played normally. This could be due to one of two factors; music is not bidirectional like language is or our ears are not accustomed to music being played the other way around. Most people will likely vote for the former explanation. But artists might object to the idea that music has to be defined in a certain way (literally speaking, too). One day, John Lennon—a British musician—heard his wife playing Beethoven's Moonlight Sonata on the piano. Then he thought about playing the chords backwards. From this experiment came a song called Because for his band, the Beatles. Classical music also often sounds like music even when it is played in reverse. In fact, an untrained ear of classical music may have difficulty guessing the "proper" direction of a piece by listening to it. The synapper might turn out it is more than just a word-parsing instrument for language.

A Nutty Theory

If language and music have the same underlining structure, then could it be also true for numbers and mathematics? Unfortunately, the answer seems to be a clear no for numbers. Like noun words, numbers have no syntax by themselves. But mathematical equations are a different story. Not only they contain numbers, but they also have mathematical symbols that give rise to some meaning. Here is a simple mathematical equation:

$$2 + 3 = 5$$

This equation can also be expressed in words:

Two plus three equals five.
Two and three is five.

In fact, all mathematical equations can be expressed as verbal thoughts. So both thoughts and equations can be output results of the synapper. The following shows the same equation as a synapper model:

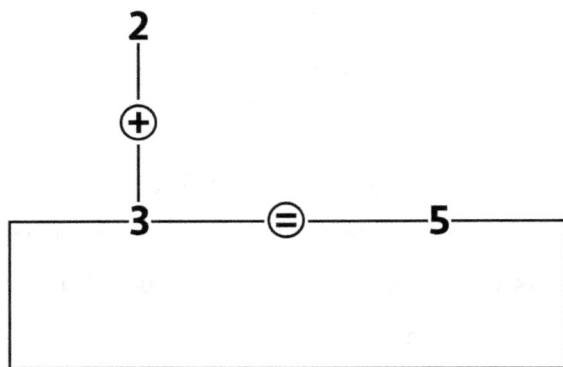

Figure 3-17. The Synapper and Math
Like verbal expressions, mathematical equations also have properties of the synapper and are bidirectional.

It is intriguing how the synapper can be applied to math the same way it applies to language. Equations such as "2 + 3 = 5" follow the same order as SVO languages:

Two plus three (S) equals (V) five (O).

Because the equal sign behaves as a be verb, switching the direction from left-to-right to right-to-left works the same:

Five (S) equals (V) two plus three (O).

Now three different forms of neurocognition (language, music, and mathematics) have been shown to possess a certain data-processing structure.

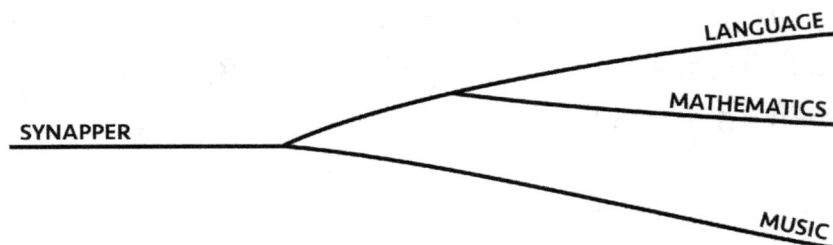

Figure 3-18. The Origin of Human Cognition
Did the synapper give our species not only language but also math and music?

This suggests a possibility that language, music, and math have a very similar underlining mechanism. However, it does not mean the three are identical triplets. People who suffer from amusia cannot comprehend music. For them, music is nothing but discomforting noise as if someone is playing random keys on a piano. But they do not have any problem with utilizing language or mathematics. On the other hand, people who have a condition known as dyscalculia may not understand math at all. But their linguistic and musical skills are intact. Like the case of Coleman described in Chapter 2, one can sing and not be able to speak. So it appears language, music, and math could be independent of each other.

What the synapper does suggest is that the three forms of neurocognition may share a common nucleus. In physics, different natural forces exist that govern our Universe; gravity, electromagnetism, and weak and strong interactions. Some physicists believe three or all four of these forces can be unified as one, a theory known as a "grand unified theory" (GUT) or a "theory of everything" (TOE). By the same token, a claim can be made that language, music, and mathematics all come from the same thing, a theory we could call a "neurocognition unification theory" (or simply NUT). It is perhaps not too far-fetched or nuts to think there is a reason why these neurocognitive traits are uniquely human and do not appear in any other species. In other words, if another species of mammals could have language, it could potentially acquire music and math as well. If that were ever to happen, we might have to reconsider what it means to be human.

Humans vs. Chimps

Thanks to Matsuzawa and his research team, we have proof humans are not necessarily the best in all cognitive areas.

Experiments show chimpanzees can outperform humans in a game of numbers in both accuracy and speed. But how is this possible? Not only Homo sapiens are equipped with bigger and more complex cerebrums, but we also have language, music, and math at our disposal. By contrast, chimpanzees do not seem to have anything special or unique about their brains. How can we humans possibly be second to such primitive creatures? According to Dora Biro, Christopher Flynn Martin, and Tetsuro Matsuzawa, ancient humans might have had to let go of some cognitive abilities to make room for something new. Once language arrived in the brain, our non-verbal cognitive skills such as analytics and visual memory likely became less important. If we can further test how intelligent chimps really are, perhaps we can learn more about our ancestors and about ourselves.

Board games such as chess and go are too complicated for chimpanzees to play since the rules must be explained in words. (In theory, chimps might be able to learn to play checkers or draughts, although it most likely is a tall order.) That is why I came up with "chimgo," a bingo-like board game with rules so easy, even chimps might be able to play. Here are the rules:

Rule 1: Player One goes first by selecting 1 on the board. On subsequent turns, she will select 3, 5, 2, 4, 1, 3, 5, 2, 4, 1, 3, 5, 2, 4...

Rule 2: After the first move made by Player One, Player Two gets to select 2 on the board. On subsequent turns, he will select 4, 1, 3, 5, 2, 4, 1, 3, 5, 2, 4, 1, 3, 5...

Rule 3: Like bingo, the first person or chimp to have five numbers in a row wins the game. The five numbers must form a straight line horizontally, vertically, or diagonally.

Rule 4: If no player has five numbers in a row after the 25th move has been made, the game ends in a draw.

Figure 3-19. Three Chimgo Boards
Numbers are placed randomly at the beginning of each game.

Since it is already a proven fact that chimpanzees can learn to count, it will be possible to train them to connect five numbers in a row to receive some reward. With repetition, chimps should be able to understand the rules fairly easily. Here are some examples of chimgo game results:

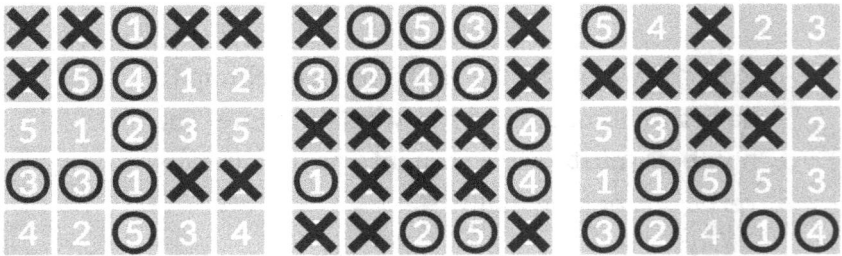

Figure 3-20. Outcomes of Three Chimgo Games
Game 1 was won by Player One (O). Game 2 ended in a draw. Game 3 was won by Player Two (X).

If a human and a trained chimp were to play this game, it is more than probable that the chimp will win at least some of the time (especially if a time limit is imposed). Depending on how well chimpanzees play this board game, we can then analyze what this might mean for our species. If it turns out chimps are clearly better at playing chimgo than most humans are, then we can speculate how much non-verbal brain power we had to sacrifice to become chatterboxes. The more our brains have changed, the more likely language was such a potent advantage for our species. Losing to chimps at chimgo might be the price we have to pay for having language.

Aliens Say the Darnedest Things

We consider ourselves to be the most intelligent among all the members on Earth. At the same time, we believe an alien species from a distant place is likely much smarter than we are. Unless aliens come to our planet on a highly-advanced spaceship, there is no telling whether it is true or not. To reach our level of cognition, the aliens would have to possess some type of language that is similar to ours. But the likelihood of them being able to speak human languages seems like one in a million. For an alien species to have language like Homo sapiens, they might have to evolve in a way we did. We were an arboreal species that later became semiaquatic and then carnivorous hunter-gatherers. Along the timeline, we possessed language, music, and math.

Unless similar conditions arose for the alien species as well, the chances are they will not speak or think like us. Even if they have language, they could use a completely different modality to communicate (e.g. optical pulses of light). Nouns and verbs may not exist in their language's parts of speech or they could have 25 different types of nouns and verbs. Recursion might be different or entirely missing from their language. All of these possibilities of alien language might give us a glimpse of whether we will ever run into them or not (assuming they exist). The way extraterrestrials think and talk, if they do, will be a good indicator for measuring the degree of their intelligence.

So let's think about aliens arriving on Earth in the near future. Since they have developed the technology to transport themselves to other places in the Universe, it is safe to assume they are a more advanced species and superior and more intelligent than we are. That means they must be brilliant scientists and mathematicians. They could have more than just language, music, and math, giving them a better chance to harness the natural forces. (They also might have unified all the natural forces into one, which would allow them to control gravity and more.) It is likely either their language and mathematics are far more complex than ours. In addition, their synapper—or the equivalent of it—might be in a higher dimension, giving them more of cognitive complexity and better odds at understanding the Universe.

If we ever get to engage ourselves in a conversation with foreigners from outside the solar system, it is probably because they are not drastically different from us. For them to have thought, they must be intelligent. For them to have communication, they would have to be social creatures. For them to have highly-advanced technology, they likely have understanding of mathematics at a level beyond ours. Still, it is not very clear why they would make such a long voyage to our planet in the first place. Judging from our own history of intercontinental travel, their purpose of visit likely falls into one of the following categories: environmental change, exile/ escape, exploration, trade, colonization, and war. Depending on the answer, we could be in some big trouble. Fortunately, because all the stars are light-years away from Earth, the chances are they will be long dead before they can make an entrance. The best bet is they will likely stay home and send out electromagnetically-coded messages to see if anyone responds. Once we are able to decipher their language, then we could say howdy and become their intergalactic penpals.

Designing Superlanguage

If it is indeed possible for aliens to have more complex language than our own, then we might be able to come up with something similar ourselves. The only problem is that we do not have innateness for this artificial superlanguage, which means fluency will be hard to attain, if not impossible. But having such superlanguage can have considerable benefits nonetheless. Potentially, it could turn us into an alien species with advanced spaceships. We might be able to take a trip around our Milky Way or even travel to other galaxies like Andromeda.

To create superlanguage, introducing new and vastly different parts of speech or word classes may be required. That means we would need more than just the synapper. Without another mutation, our brains will not be able to utilize superlanguage naturally. But, as genetic engineering and genome editing become more commonplace in the future, it could just be a self-induced genetic modification. Language might receive an upgrade the same way we have done it with mathematics over the years. It is truly remarkable how a method of making simple additions and subtractions turned into differential equations that tell us the secrets of the Universe. If such progression is possible for math, language also might evolve into something more.

For instance, a sentence such as "The party starts at nine" is a rigid statement and does not have much room for variation. The subject and the object in the sentence are both constants. But math equations like the following are quite the opposite:

$$F = ma$$

Also known as Newton's second law of motion, this formula has three variables: force (F), mass (m), and acceleration (a). Since their values can be anything, the expression shows a relationship between these concepts rather than making a firm statement about one particular situation. By including variables into verbal thoughts, language can show fluid relationships between subjects and objects. As an example, we can take the following sentence and derive two variables from it:

The party starts at nine.

First, we take the subject and the main verb ("The party starts") and turn them into a variable of time. Then we can define the object ("nine") as a unit time, o'clock.

$$T_{party} = 9{:}00 \text{ PM}$$

We can also change the condition so that the party will start when ten or more people show up.

$$T_{party} == N_{people} > 9$$

We can use two conjoined equal signs to represent a condition or correlation whereas one equal sign would indicate a definition. Thus, the time of the party (T_{party}) is when the number of people (N_{people}) has reached at least ten. Not only can this formulaic expression be applied to other parties, it also signifies what *party* actually means.

Here is a different example:

Bob was killed in a plane crash.

We can convert "Bob was killed" into Bob's death (D_{Bob}). A "plane crash" can be redefined as the number of fatal malfunctions of a plane (FM_{plane}) such as collisions and explosions. Therefore:

$$D_{Bob} == FM_{plane} = 1$$

We can see that Bob's death is related to the number of fatal malfunctions of the plane. But if it turns out the news is wrong and there was no malfunction, then the value of FM_{plane} will change to zero and the state of D_{Bob} will also be affected.

Here is another sentence:

Emily has never heard of Nirvana.

Since the memory of Emily (M_{Emily}) has no knowledge of Nirvana ($K_{Nirvana}$), the sentence can be expressed as the following:

$$M_{Emily} == K_{Nirvana} = 0$$

As the amount of $K_{Nirvana}$ increases, we can then guess whether Emily is a fan of the rock band or not.

Here is one more example:

The professor was late to class by 20 minutes.

We can form a relationship between the professor's time of arrival ($ToA_{professor}$) and when the class is scheduled to begin (T_{class}).

$$\text{ToA}_{\text{professor}} = \text{T}_{\text{class}} + 20 \text{ min.}$$

Suppose the professor arrived right on time. Then the relationship between the two variables would be:

$$\text{ToA}_{\text{professor}} = \text{T}_{\text{class}}$$

In other words, the class will not start unless the professor shows up. These expressions transform ordinary verbal statements into something more; conditional relationships between subjects and objects as variables. Numbers become significantly more important in this type of a hybrid between math and language. If this were the natural state of our thought, we could perhaps be more logical and be better at discovering the truth in every situation. Unfortunately, humans did not evolve to speak a math-language hybrid. We can still perform this skill to a degree in mathematics and computer programming, but it feels as if we are speaking a foreign language. If early humans were so brilliant they invented language, then it should not be a problem for modern humans to exercise this form of communication into their daily conversations. However, Noam Chomsky might say:

Probability$_{this}$ = 0

The Verdict on Chomsky

Emerson Pugh, a computer scientist, once said the following (which I call the "complexity paradox"):

> If the human brain were so simple that we could understand it, we would be so simple that we couldn't.

So either we are not smart enough or the human brain is just way too complex for anyone to figure out. (Both can be true as well.) No matter what the situation may be, we cannot examine our minds to the fullest if Pugh is right. The only chance we have is to create an independent machine of intelligence that will study the human brain for us. Of course, inventing this machine would not be such a challenge if we were intelligent enough in the first place.

Noam Chomsky is one of the early ones to take on this assignment. From an early age, he chose language as his field of research and no one has influenced linguistics more ever since. Even the word *linguistics* was not a popular term before Chomsky became a linguist. As much as his name has been synonymous with the study of language, criticism of Chomsky also grew steadily over time. His critics do not hesitate to refute Chomsky's linguistic claims likely because his ideas seem way too theoretical and unmeaningful to most. People also have criticized Chomsky's unwavering and blunt dismissal of theories that reject his own. Since Chomsky is not so much concerned with providing evidence for his conjectures and him also not wanting to carry out any experiments gave him a reputation of being "unscientific" by some. But, perhaps more than anything else, his progressive political ideology has garnered enemies from both linguistic and non-linguistic scholars.

But one thing people often neglect to do is launch a direct attack on Chomsky's views on the origin of language. When they do make an attempt to counter Chomsky's claims, they almost always fail to provide any convincing evidence and often show no congruence with what we have discovered so far. In a similar manner, people in the 17th century did not try to refute Galileo's heliocentrism with logic and reasoning. Instead, their reactions to Galileo's work were quite coarse and not based on anything scientific. Sadly, Galileo died without ever seeing real imagery of Earth.

For the purpose of this book, we should not forget the most important element of our discourse; is Chomsky right about language? Is language really innate? Are all human languages the same in essence? Was it a genetic mutation that gave us language? The answers to these questions will ultimately cement his legacy. In the meantime, we are free to believe whatever we choose to believe, just like the old days when Galileo was thought to be wrong.

THE FUTURE

"The greatest souls are capable of
the greatest vices as well as of the
greatest virtues."

René Descartes

We have discussed so far how language might have emerged
in our species tens of thousands years ago. But digging up
the past could also give us an insight into knowing what lies
ahead. Once we figure out how and why we became talking
bipeds, we may be able to predict what will come of Homo
sapiens in the distant future. Since language makes up a
large portion of the reason for our current status as the most
advanced species on Earth, we can suspect what it will do to
us as we continue to evolve. Because of how the word *evolve*
is generally understood and used in context, many people
might assume humans are genetically progressing in some
way. But, if evolution has taught us anything, the act of

evolving does not necessarily result in progress. Humongous dinosaurs evolved into tiny birds so they could survive and proliferate in their rapidly-altered environment. Evolution is—more or less—change without a direction. A species is not supposed to get better at anything as it evolves except at adaptation. Concepts such as natural selection and survival of the fittest might give off an aroma that Mother Nature only chooses individuals with the "best genes" for the next generation. But this is hardly true. If anything, the objective of evolution is to proliferate most genes as much as possible, a strategy of having quantity over quality. For a male salamander to spread his genes, it may not be such a good idea to be choosy when it comes to finding mating partners. He should mate with as many female salamanders as possible, not caring whether they possess good genes or not. To have genetic diversity, a species is more likely to implement a policy of "extinction of the unfittest" rather than a policy of "survival of the fittest." Even though they may sound alike, the former policy gets rid of the worst genes (e.g. the bottom 20%) whereas the latter policy only selects the best genes (e.g. the top 20%). For our species, it was likely mutation and genetic drift that caused language to spread. Once the language mutation took place in one or a few individuals, it slowly became fixed in our population over time. But why and why only us? We do not know the answer yet. Although the cause might have been just a small tweak in our DNA, the effect cannot be more astonishing. It is rather rare that a species experiences such enormous overhaul so suddenly. More than likely, our environment went through a drastic transformation and many individuals lost their lives as a result. This means humans will likely encounter another major genetic modificaiton in case of a catastrophic event.

The New Homo

According to research studies, dyslexia and dyscalculia are mostly caused by genetic factors, making them hereditary. It is unclear which specific genes are associated with these neurological disorders. But some scientists believe both could be caused by the same genes in some instances. Dyslexic children in China may have slightly different intraparietal sulci (grooves on the surface of the parietal lobe). Usually, this particular region in the brain is the cause of dyscalculia and not dyslexia. But Chinese characters are visual representations like Egyptian hieroglyphs. So any problem with analyzing the visual aspect of written symbols can make someone both dyscalculic and dyslexic. Even though dyslexia and dyscalculia generally do not affect one's speech, they still end up being major hurdles in his life. People who are dyslexic struggle to read written words. People who are dyscalculic have trouble with reading the analog clock, being on time, shopping, and many other activities. They have almost no number sense and can only count with their fingers even when they become adults.

It is quite interesting how persistent these traits remain in our population when almost nobody has any difficulty with language in its natural forms. The fact that neurological conditions such as dyslexia are hereditary suggests they may not disappear any time soon. Various reports show about five to ten percent of people have either dyslexia or dyscalculia. But the number may be on the rise. Both dyslexia and dyscalculia are considered dominant traits. So if one of the parents is a carrier, the child has a 50% chance of having the same trait. Assuming people with dyslexia or dyscalculia have the same number of children as others do on average, this will not affect the number to rise at all. Instead, people who are absent of these traits could make dyslexia or dyscalculia more prevalent. It is because, due to many factors, some of their offsprings can be born with these conditions as well. One of the factors might actually turn out to be language.

As we treat language as a necessity of being human, our dependence of it will not waver anytime soon. It is language that makes us so unique and special. Everything else— including music and mathematics—is secondary to having an ability to think and speak. Even if someone does not comprehend music or math, it does not matter all that much in the grand scheme of things. The person will still be able to live a normal life for the most part. But it is difficult to even fathom what it is like to live as a human being without language. Over time, we have decided to associate ourselves closely with people who use words. As a consequence, the language gene became fixed in every human being. At the same time, almost all other cognitive abilities have become expendable. It might seem as if nothing much has changed for our brains since the arrival of language. But, as we continue to make a push for our linguistic trait even further, it may be just a matter of time before we slowly evolve away from numbers and math.

The Days Are Numbered

Two mathematicians are in a taxicab numbered 1729. One mathematician sees the number and comments to the other about how dull it is. The other mathematician responses by saying it is not really a boring number. He says 1729 is the smallest number that is the sum of two different numbers that are cubed in at least two different ways.

$$1^3 + 12^3 = 1729$$
$$9^3 + 10^3 = 1729$$

Any number smaller than 1729 does not meet this requisite. It was Srinivasa Ramanujan who told G. H. Hardy this fact in 1919. It is uncertain whether Ramanujan, a mathematician from India, already had thought about 1729 beforehand or he simply worked out the number upon seeing it. Either way, there is little doubt he was a one of a kind. Before becoming sick and passing away in his early 30s, Ramanujan published numerous ground-breaking mathematical theorems that shocked other mathematicians around the world. A century later, India is still in search of someone that could rival Ramanujan's mathematical ingenuity. After his death, India has experienced an exponential population growth and is about to become the most populous nation in the world. But the next Ramanujan still has not been found. After his death in 1920, over a billion people were born in India. Outside of India is a man named Grigori Perelman, a Russian mathematician born in 1966. Perelman proved the Poincaré conjecture, which was considered one of the most difficult problems to solve in the field of mathematics.

The truth is, a great mathematician is almost impossible to find. After the birth of Euclid around 300 BC, Fibonacci (1175), René Descartes (1596), Leonhard Euler (1707), Carl Friedrich Gauss (1777), Carl Gustav Jacob Jacobi (1804), David Hilbert (1862), Srinivasa Ramanujan (1887), Paul Erdős (1913), and Grigori Perelman (1966) arrived and became some of the best mathematicians in history. The world would be a completely different place if these figures did not exist. They gave us a much deeper understanding of Nature, resulting in the development of advanced technology that is way beyond what our early ancestors had. And yet, these genius minds are needles in a haystack. Why is that everyone can utilize language but most humans do not have a good grasp of higher mathematics? Why are people like Ramanujan so few and far between? Is it possible we are slowly losing our natural analytical skills as a species? Unfortunately, the answer might be yes.

There are two reasons why our brains could be becoming less and less mathematical. The first one is language. The human brain is no longer growing in size; it actually seems to be getting smaller. If we were to maintain our ability to utilize language, brain regions related to other cognitive functions may have to be downsized. The fact that chimpanzees can outperform humans at a number-memory game suggests we might have already lost some analytical or memory-related abilities. The second reason has to do with modern society. Prior to the 20th century, most people got married by the time they were in their early 20s. They had multiple children and had similar life styles. But things started to change at the end of the 20th century. According to a New York Times article written by Claire Cain Miller, many women in the United States in 1980 had their first babies when they were around 18 to 21 years old. In 2016, almost half of the women that got pregnant for the first time were between 25 and 33 years old, much older than the other half. An age gap began separating people into two groups. In the early 21st century, women in the U.S. without college degrees have their first pregnancies when they are around 24 years old. However, that number increases to 30 for women who graduate from college. The reason for this gap turns out to be difference in the level of education but also where they live. On average, women living in big cities such as Manhattan and San Francisco become first-time mothers after they turn thirty. In rural areas, women are still in their early 20s when they get pregnant for the first time. If people with above average analytical skills spend more years in higher education and therefore wait longer until they start having babies, then it would take them more years to proliferate and produce fewer babies compared to others. (The big assumption here is people without above average analytical skills will not spend as many years in higher education, partially because they have fewer opportunities to do so.) In other words, not only do highly-educated couples take more years to have babies, but they also have fewer babies than the rest. This will likely lead to only one outcome; the percentage of mathematical brains in our species will continue to diminish over time. As our brains keep getting smaller and smaller, this might be our inevitable fate.

Our Destiny in Laughter

So far, we have dealt with nothing but depressing news about our potential future. But is there any upside to the way we might evolve? When something is lost, usually something else is gained. As we slowly phase out our primitive traits of analytics and mathematics from our brains, we might gain superiority in something else. Memory and music are viable candidates, but language could also receive an upgrade. It is our ability to talk to others that makes us want to socialize with people. Homo sapiens is a social species that is designed to long for contact with other members emotionally, physically, and verbally. So we have no choice but to hold on to language for dear life. This could mean we will have better teachers, better speakers, better interpreters, better lawyers, better storytellers, and funnier comedians in the future. While our analytical skills go down, our verbal skills might travel in the opposite direction. Hundreds of generations later, people might speak five different languages as a bare minimum, including Americans. (However, it is uncertain how many languages will remain at that time.) The question is, how long will it take for us to reach that point?

Suppose we are losing 0.1% of our non-verbal cognitive functions per generation. Assuming women give birth to their first babies at the age of 25 on average, the "reproduction cycle period" would be 25 years. Under these circumstances, it will take approximately 25,000 years for us to become completely non-mathematical. If the rate is more like 0.5%, it would not require more than 5,000 years. By this time, we may not have any scientists, engineers, and mathematicians. Additionally, bankers, accountants, stockbrokers, doctors, and nurses also might struggle with their jobs. Athletes will be among them as well since analytics is a key part of playing sports. Any job that requires abstract or analytical thinking may not be filled, at least not by humans.

But we happen to be a species of perseverance. As long as we realize the potential decline of our intelligence in coming centuries and millennia, we might be able to change our own destiny. (Sadly, seeing our lack of response to rapid climate change somewhat proves there may not be much intelligence left to save already.) To have some hope of what may lie ahead for our species, we can consider two things; genetic engineering and artificial intelligence. If we know which genes are responsible for analytics, then we will be able to introduce them back into the population with genome editing. The other solution is to develop artificial intelligence for the sole purpose of saving humanity.

Heaven or Havoc?

In 1956, a group of mathematicians and scientists got together to conduct research and share their ideas on artificial intelligence. They thought they could program machines to learn and think like humans. Language was listed as one of the properties machines have to possess in order for them to simulate our behavior. This gathering, known as A Proposal for the Dartmouth Summer Research Project on Artificial Intelligence (1955), stated the following:

> 2) How Can a Computer be Programmed to Use a Language
>
> It may be speculated that a large part of human thought consists of manipulating words according to rules of reasoning and rules of conjecture. From this point of view, forming a generalization consists of admitting a new word and some rules whereby sentences containing it imply and are implied by

others. This idea has never been very precisely formulated nor have examples been worked out.

The attendees generally agreed it would only be a matter of a few decades, if not sooner, before machines start thinking like humans. More than six decades later, almost all of the attendees have now passed away. The artificial intelligence they once envisioned still has not arrived. What we have so far are machines that can play board games and do a few other cognitive tasks, although the list seems to be growing fast. They can translate simple sentences from one language to another in real time. They can even understand human speech and respond in words through the works of complex algorithms and having massive amounts of data. But no machine yet thinks and learns like humans. Perhaps it should not be such a surprise. How can we possibly build a machine to think like us when we do not even know how we think? Before we invent true artificial intelligence that can utilize human language, first we must figure out what human language is and how it really works. The quest of finding the origin of language, therefore, can have some serious implications. What if we actually create conscious beings that can think on their own? How intelligent will they be one day? What can they do for our future?

AI machines we are anticipating are what could be described as first-level artificial intelligence. The machines will be able to learn and think independently like humans, but they will not be equipped with any emotions. Also, they will not be able to reproduce like biological organisms do, making them unable to evolve on their own. Only humans are able to design, modify, and terminate them. Therefore, the first-generation thinking machine will not pose a major threat to our species. Instead, it will likely make our daily lives more convenient and efficient. More importantly, first-level AI can potentially save millions of lives by assisting doctors and surgeons, developing better drugs, and reducing the number of fatalities in our daily endeavors. Improving weather forecasts by ten percentage points, say from 80% to 90%, could also be a palpable benefit for the humankind. Of course, not every consequence will be constructive. A major concern of having artificial intelligence is that it will most certainly be exploited in war. Whether human casualties will see an increase or decrease due to AI's existence is anyone's guess.

When artificial intelligence reaches the next stage, it will become much more self-governing. Following orders from humans is no longer a necessity. Not only will machines think independently like humans do, but they also might possess intangibles such as love, hatred, motivation, fear, desire, irrationality, and unpredictability. Without the implement of these features, they will lack social intelligence and may not be fully conscious. Machines with second-level AI will be able to create new versions of themselves on their own. At the same time, they will still be under the control of humans. We can still decide the fate of artificial intelligence until it reaches the final stage.

Third-level artificial intelligence will be a different story. With AI in its complete and final form, we will not have much choice but to relinquish what was once our beloved creation. By then, machines will become too powerful and ubiquitous to be placed under our control. AI might be interwoven into our daily lives so much that we would simply not be able to live without it. The distinction between artificial intelligence and natural intelligence will not be so discernible in the future when bionic chip implants become as popular as having a calculator. Everything, including the world economy, will either directly or indirectly be depended on artificial intelligence. Instead of humans and AI struggling to coexist, the two parties will likely develop a relationship that will be beneficial to both. Artificial intelligence will simply become an inseparable part of our lives. The only concern is, while machines continue to improve themselves, humans will have no choice but to just sit and watch. It is unclear what events will take place once we are no longer an integral part of AI's existence. Potentially, this could be the beginning of the end for our species.

As despondent as it may sound, we should also be more objective of what awaits us in the next 5,000 years. With a rising number of nuclear weapons around the world, one of the greatest threats for our species will always be war. It would only take a handful of political leaders to decide the exact date of our demise. But there are other major problems as well. As the standard of living increases in many developing countries, people will compete aggressively over natural resources. Even basic needs such as food, water, and shelter may become scarce globally. Climate change will make many regions around the world uninhabitable. But how are we suppose to solve these issues all by ourselves? What are the chances that people will become much wiser in the near future to make them disappear?

We must come to terms with the reality that humans will face extinction some day. We cannot possibly go on for millions of years when our needs have skyrocketed like our population. It is unknown whether we will still exist 500,000 years from now. By this time, Earth will most certainly have gone through major transformations, making life extremely difficult to persevere. It is possible we might colonize other planets so billions of people can live far away from Earth. But the most likely scenario is our species will cease to exist before we can develop the technology to visit our relatives on Europa or Enceladus (Jupiter's and Saturn's moons, respectively). If the chances of total annihilation of our species increases incrementally by 0.0005% every year, then we will be all gone within 200,000 years or so. The only question is, are we approaching our doomsday by mere 0.0005% per year? Even the most optimistic futurists would hesitate to say yes. With artificial intelligence, we might have a shot at solving most of our ongoing problems, very likely extending the lifeline of Homo sapiens. Although it may not be such a great plot for a movie, it will have to do.

Creating the Next Computer

For the machine to emulate human behavior at any meaningful capacity, language acquisition is a must. Either it has to learn language in baby steps like humans or it will be built with language from its genesis. Whether this can be achieved with a traditional computer is not all that clear. The computer was originally an invention created for handling numbers. On and off switches are used to generate binary numbers made up of zeros and ones, which turn into decimal digits such as 47 and 92. The computer's brain, commonly known as the CPU, compares two different numbers (input) and makes a computation (output). Since math is supposed to be exact, the same input always produces the same output. Adding five and nine should not result in anything other than 14. No other outcome is allowed. The computer is supposed to be nothing but clockwork. This precision design works extremely well with numbers but not as much when it comes to language. Asking what 5 + 9 is to young children in kindergarten could produce a plethora of answers such as "Fourteen," "One four," "I don't know," "I ran out of fingers," and "I think I peed my pants."

Computers do not think and talk like humans because not only they lack language but they are also missing consciousness and unconsciousness. In other words, they have no means of expression and they have nothing to express. The computer is more like a plant or a fungus than it is like an animal. It is just an automaton that has no need for thought. But if that need can be fulfilled by humans, then the only missing ingredient is language. For the machine to utilize language, it should have a circuitry similar to the language faculty found in the human brain. If not, the machine is only giving the appearance of speaking when it is doing something completely different inside the circuitry. Imagine a human character in a video game walking. As he wanders around on the screen, his legs move back and forth. But this is only an illusion since the game is simply shifting his entire body from point A to point B. The only reason why the game developer decided to move the character's legs back and forth is to create a sense of realism. Comparably, for the machine to truly talk, it must resemble the human brain. For it to resemble the human brain, it must have the synapper.

The central processing unit (CPU) for the synapper will be equipped with a similar modern computer architecture. But, instead of having an arithmetic logic unit for dealing with numbers, a language processing unit will take the center stage. It will include an expression analysis unit that can decipher verbal expressions. This architecture will be able to generate responses to the received input based on conditions such as the machine's database or memory. This "computer" does not actually compute numbers but understands language and responds in words. The responder will be the language faculty for the machine.

Applications for the responder is virtually limitless. For one, it can be implemented as a bionic chip for the brain that can restore any loss of language functions, similar to what bionic eyes can do for the blind. In addition, the responder can be used to deliver the news, translate one language into another, teach, and do research. It might even be able to do one's homework. However, the responder still does not make the machine conscious. That means it will not be able to think independently by itself or make its own decisions. Nonetheless, the invention of the responder will be a major step forward in creating our AI companions.

The Future of Linguistics

It comes as a surprise how little people care about what language is and how we got it in the first place. Most curious minds would rather explore the outer space or learn about dinosaurs that predominated on Earth more than 50 million years before our earliest ancestors roamed around. It does not seem to bother almost anyone that we know little to nothing about the cause that gave our species such mental acumen, language or otherwise. Even the very few who have shown interest in search of the origin of language mostly have moved on. They argue the topic is impossibly difficult; it is nothing more than a waste of time. However, rapid development of neuroscience and genetics might change all that in the very near future. By 2050, we may find out how we became thinking apes. More importantly, we could also discover the true nature of our species and what we really are.

Be that as it may, a few people did make attempts to study what language is and how it works. Ferdinand de Saussure of Switzerland took on the challenge in the late 19th century and the early 20th century before his death. Saussure's work laid the foundation for what eventually became modern linguistics. The study of language in the future might belong to neuroscientists and neurolinguists. Traditional linguists will have no choice but to evolve if they want to avoid extinction. As first-level artificial intelligence gets off the ground, linguists may have to spend more time dealing with machines than with humans.

With great power comes great responsibility. As we gain more understanding of ourselves, we also gain the power to create something we have never seen before; something that can make our lives better or worse.

Depending on the decisions we make as we move forward, the fate of our future is ultimately in our hands. Let us not forget famine, pollution, war, injustice, discrimination, and corruption are all our own creations. To create a better world, we must seek for knowledge that will lead us to wisdom. If the truth is to set us free, then we shall welcome the truth of whatever it is that makes us human.

APPENDIX

Synapper Formation

To put a sentence into its proper synapper structure, one must first determine the word type for every word in the sentence. Usually, sentences will include both nodes and branches, although some might not have any branch words. This procedure heavily depends on words' parts of speech. Since each language's lexical categories are slightly different, the rules of synapper formation will also vary between languages.

In general, it appears that dependent words become branches to independent words. By contrast, independent words cannot be branches and are always nodes. For English in particular, the following rules can tell us which words are branches:

[1] A noun if the next word is also a noun
[2] An adjective if the next word is a noun, a pronoun, or another adjective
[3] An adverb if the next word is an adjective or another adverb
[4] A determiner

These rules do not apply to the last word of a sentence, which is always a node. Once the branches have been labeled, all the other words in the sentence can then be labeled as nodes. The most simplistic and practical style of synapper modeling will have all branch words attached to nodes in one perpendicular direction from the stem.

But some words such as *reach* and *fast* belong to more than just one part of speech. For humans, this is usually not an issue since we naturally know what each word's correct part of speech is in the given context. But this is not the case with machines. So computer software has to implement many supplementary rules in order to generate the proper synapper structure.

Sentences

A sentence is a set of words that represents a thought, a request/command/order, or the emotional state of an entity. Usually, a sentence has to possess some degree of syntax for it to convey meaning effectively. It can also be used for purposes of communication and socialization.

A sentence is comprised of one or more words. For a sentence to represent an independent thought, it requires a subject and a verb. Here are examples:

Birds fly.
They fly high in the sky.

The first sentence is an independent thought. It does not depend on any other sentence to convey its meaning. However, the second sentence is a dependent thought because of the pronoun *they*. Without the first or another sentence, the subject of the second sentence cannot be determined. (For dependent thoughts like the second sentence, sometimes the subject can be omitted in some languages like Korean.)

In theory, a sentence can have an infinite number of words. Also, all natural languages are known to be capable of having recursion. This means an infinite number of sentences can be produced even with only a few words such as *Tom, Sally, and,* and *jumped* like the following:

Tom jumped.
Tom jumped and Sally jumped.
Tom jumped and Sally jumped and Tom jumped.

The problem with representing a thought as a written sentence is that the sentence becomes more and more complicated to analyze as the number of words increase. This is because words are all written in the same direction. With synapper modeling, the structure of a sentence becomes a lot more intuitive when multiple directions or dimensions are used. Here is an example:

The brown bird flew over a big tree.

(The)-(brown)-<bird>---<flew>---<over>---(a)-(big)-<tree>

( ) = branch, < > = node

This style of writing might make reading easier for some people, especially if they are dyslexic. In addition, removing branches of a sentence will yield a core sentence that is only composed of nodes. This simplified form of a sentence shows the essence of its original meaning.

<Bird>---<flew>---<over>---<tree>

By removing the branches, the previous sentence has become one-dimensional, which is easier to understand and translate. Core sentences remain grammatically intact even when branches are gone.

Words

Words are units of association used for accessing certain neural activities in the brain. For instance, the word *dinosaur* can regenerate imagery of a giant reptile roaring in the wild. Without words, it is far more difficult to recall neural connections representing a certain object, event, or concept stored in the brain's memory. Words are the building blocks of sentences and therefore the building blocks of thoughts. Language makes use of words and affixes in a systemic way to convey meaning.

It may seem like words are inherently sound-based, but this turns out to be not true. Deaf people, for example, use hand signs as words without involving sound. Even written text can be considered words although this is somewhat different from vocal sounds and hand signs. For people whose native language is an oral language such as Portuguese and Malay, written forms of words are directly linked to words and not what words actually symbolize. For instance, the utterance of the word *dinosaur* ("die-nuh-sore") is a first-level word representing an image of the gigantic animal. By contrast, the written word *d-i-n-o-s-a-u-r* represents the sound of the word and not the concept itself. Thus, written words should be considered second-level words that are based on first-level words.

First-level word:
"die-nuh-sore" (sound) -> dinosaur (concept)

Second-level word:
d-i-n-o-s-a-u-r (text) -> "die-nuh-sore" (sound)

Instead of text being representations of first-level words, it can also become words in the first level, representing concepts directly without having to involve another modality. One way of achieving this is by using six-digit numbers as first-level words. (With six digits of numbers, up to 1,000,000 words can exist, which should be more than sufficient.) This method also eliminates the dependence of belonging to a particular natural language like English. An extra digit is attached in the beginning and at the end of a six-digit number to function as a prefix and a suffix. Here is an example:

0|000196|0

This first-level word represents the concept *house*. Since it is without any affixes, the first and last digits are zero. Here is another word:

0|000528|3

This word, 00005283, means *jump*. But the suffix is 3, indicating the word is in the past tense, *jumped*. These first-level words can be used with synapper modeling, giving visual representations of thoughts. Here is what the proverb–*All work and no play makes Jack a dull boy*–looks like as a synapper-structured sentence:

(All)-<work>---<and>---
(no)-<play>---<makes>---
<Jack>---(a)-(dull)-<boy>

(00004290)-<00007920>---<00000280>---
(00000730)-<00005840>---<00008691>---
<00520480>---(00000110)-(00030820)-<00000410>

The first synapper model is clearly in English and each word represents a certain sound. The second model, however, does not belong to any particular language and the words represent concepts instead of representing first-level words. From this model, a thought can be expressed in any natural language by replacing each number with a word with the same meaning in a designated language. This method brings us one step closer to simulating the process of generating individual thoughts in the human brain. Another advantage of this method is that sentences constructed in this manner can be translated into any natural language without error. Here is an example:

<Did>---<you>---<say>---<orange>

<00000363>---<00000050>---<00000660>---<00007180>

The word *orange* in the first sentence can mean either the color or the fruit. But with first-level words like in the second sentence, there is zero ambiguity of what each word means since the word *orange* is given two different words (or numbers) for representing two distinct concepts. For instance, the translation of 00007180 in Spanish is either *anaranjado* or *naranja*, the former being the color and the latter being the fruit. But it cannot be both.

Defining Words

Defining a word is a process of forming one or more associations between a neural connection generated from some sensory input and a unit of a modality. When a child sees a dog, some neural activities take place in his brains. His mother pointing to the animal and saying the word *dog* out loud will form an association between the two for the child. If humans do not have sensory organs like eyes and ears, new words cannot be learned and therefore language becomes useless. Helen Keller, who lost sight and hearing at a very early age, eventually learned to utilize language through her sense of touch. Anne Sullivan, her teacher, would spell out words on the palm of her hand. Using this technique, tactile words gave Keller a way to generate thoughts like any other person with language.

But not all words are created with sensory organs. Some words seem to be preinstalled from the beginning. Words such as *or, beyond, nevertheless,* and *presumably* are not learned by seeing or hearing. Noam Chomsky asserts some concepts are inherently human; although some of these concepts are also observed in other mammals. Language would not work the same way unless both intrinsic and acquired concepts exist. This rule applies to machines as well. Teaching a machine first-level words that are based on sensory input has already been proven to be feasible. However, how does one teach a machine the meaning of the word *supposedly*? This task is as difficult as making a blind man understand colors. (Simply giving a definition of the word will not work since it is not a concept by association like the word *umbrella*.) Ultimately, it may not be possible to create a form of artificial intelligence that can think fully like humans without duplicating our consciousness and unconsciousness.

ACKNOWLEDGMENTS

It would not have been possible to write this book without the works of linguists, mathematicians, and scientists who paved the way for millennia. I would personally like to thank Emmanuel Keller and Frans de Waal for making contributions to the book. Another person I'd like to thank is Noam Chomsky, who has seen the light before many of us. Without his influence and diligence to the field of linguistics, this book may not have been written.

LAST WORDS

This book was written entirely by the author while being naturally alive. For the purpose of writing this book, no assistance was given by any form of artificial intelligence for creating its content. All of the written words are his except for quotes.

INDEX

4.0 The Future

4.1 The New Homo

4.2 The Days Are Numbered

4.3 Our Destiny in Laughter

4.4 Heaven or Havoc?

4.5 Creating the Next Computer

4.6 The Future of Linguistics

www.ingramcontent.com/pod-product-compliance
Lightning Source LLC
Chambersburg PA
CBHW060846280326
41934CB00007B/935